Springer Tracts in Modern Physics

Volume 240

For further volumes:
http://www.springer.com/series/426

Springer Tracts in Modern Physics

Springer Tracts in Modern Physics provides comprehensive and critical reviews of topics of current interest in physics. The following fields are emphasized: elementary particle physics, solid-state physics, complex systems, and fundamental astrophysics.

Suitable reviews of other fields can also be accepted. The editors encourage prospective authors to correspond with them in advance of submitting a manuscript. For reviews of topics belonging to the above mentioned fields, they should address the responsible editor, otherwise the managing editor.

See also springer.com

Mathias Getzlaff

Surface Magnetism

Correlation of Structural,
Electronic and Chemical Properties with
Magnetic Behavior

 Springer

Prof. Dr. Mathias Getzlaff
Universität Düsseldorf
Institut für Angewandte Physik
Universitätsstr. 1
40225 Düsseldorf
Germany
e-mail: getzlaff@uni-duesseldorf.de

ISSN 0081-3869

ISBN 978-3-642-26506-8

DOI 10.1007/978-3-642-14189-8

e-ISSN 1615-0430

ISBN 978-3-642-14189-8 (eBook)

Springer Heidelberg Dordrecht London New York

Cover design: eStudio Calamar, Berlin/Figueres

Printed on acid-free paper

Springer is part of Springer Science+Business Media (www.springer.com)

To Sarah and Tim

Preface

This book is intended to give an introduction and a comprehensive overview concerning the main areas of surface magnetism with special emphasis on rare earth metals.

Investigations in this field require experimental techniques which are sensitive to the topmost layers on the one hand and simultaneously to magnetic properties on the other hand. Using additionally tools with a high lateral resolution the visualization of magnetic domains becomes possible.

The understanding of magnetic and electronic behavior requires the knowledge of the structure on a microscopic scale. Due to this important relationship the dependence of electronic on structural properties is the first topic. This contains investigations not only on rare earth metals but additionally on $3d$ ferromagnetic systems.

It is important to keep in mind that the chemical behavior of a surface determines the surface electronic properties. Thus, variations, e.g. due to adsorbate atoms, have a significant influence. This aspect will be focused on as the next topic with the description of selected substrate layers which were exposed to different types of gaseous molecules.

Investigations on the surface magnetism of itinerant ferromagnetic materials, including the influence of adsorbates on surface magnetic properties, and magnets with localized moments is the final and main topic of this volume. It will end with the realization of laterally resolved spin polarized vacuum tunneling which enables to image magnetic domains on the nanometer scale.

Acknowledgements

This work summarizes my research on the above-mentioned topics performed at the Universities of Bielefeld, Mainz, Hamburg, and Düsseldorf.

The realization of such an extensive work as presented here is, of course, impossible without the help of many other people. It is a pleasure to thank all of them for their collaboration, their support and their interest.

First of all, I would like to thank Prof. Dr. R. Wiesendanger for the opportunity to join his group at the Institute of Applied Physics and Microstructure Research Center at the University of Hamburg. The excellent laboratory facilities, the extensive financial outfit, and the continuous support and encouragement he provided allowed me to develop the activities on scanning tunneling microscopy to that level as described in this report. I have also to thank all other members of the scanning tunneling microscopy group of Prof. Wiesendanger. The stimulating atmosphere in the group contributed a lot to success of these activities.

Further, I thank Prof. Dr. H. Steidl from the University of Bielefeld for his support concerning the spectroscopic capability on spin polarized metastable de-excitation spectroscopy (MDS) because this unique atomic beam source opens fascinating possibilities on surface magnetism due to its extraordinary figure of merit. I would also like to thank Prof. Dr. W. Raith and Prof. Dr. G. Baum for joining the group in Bielefeld.

My thanks additionally goes to Prof. Dr. G. Schönhense from the University of Mainz. His knowledge and fruitful discussions on photoemission with its widely spread capabilities allowed me to investigate "surface magnetism" in a very intensive way.

I thank Prof. Dr. Jürgen Braun (now at the LMU München) for a lot of theoretical calculations giving a better understanding of magnetic dichroism experiments.

Especially, I would like to thank Prof. Dr. Matthias Bode (now at the University of Würzburg) for his introduction and continuous help on the field of scanning tunneling microscopy and spectroscopy. Further, I thank Prof. Dr. Astrid Pundt from the University of Göttingen for her fruitful discussions on the influence of hydrogen atoms in solids. Last but not least I thank PD Dr. Joachim Bansmann (now at Ulm University) for the close collaboration on magnetic dichroism studies.

The studies on spin polarized MDS in Bielefeld were supported by my diploma student Mike Wilhelm and Ph.D. students Dirk Egert and Peer Rappolt. The work on Heavy Fermion Systems in Mainz was connected with a lot of work done by Ulrich Kübler, Mike Wilhelm, and Bernhard Schmied who wrote his Ph.D. thesis in this topic as well as Dr. Gerhard H. Fecher. For the activities on STM in Hamburg I would like to thank André Kubetzka, René Pascal, Oswald Pietzsch, Heinz Tödter and Christoph Zarnitz for all the work they have done and the stimulating discussions.

I may thank all the people of BESSY; due to their help it was a pleasure to work in Berlin.

Last, but not least, I have to acknowledge the efforts of the mechanical and electrical workshops; scientific experimentation is impossible without them.

Düsseldorf, May 2010   Mathias Getzlaff

Contents

Acronyms

AD	Auger de-excitation
AES	Auger electron spectroscopy
AFM	Antiferromagnetic
AN	Auger neutralization
bcc	Body-centered cubic
BESSY	Berliner Elektronenspeicherring-Gesellschaft für Synchrotronstrahlung
CDAD	Circular dichroism in the angular distribution of photoelectrons
CEM	Channel-electron multiplier
DOS	Density of states
ECS	Electron capture spectroscopy
EDX	Energy disperse X-ray detection
EELS	Electron energy-loss spectroscopy
fcc	Face-centered cubic
FLAPW	Full potential linearized augmented planewave
FM	Ferromagnetic
FWHM	Full width at half maximum
GMR	Giant magnetoresistance
hcp	Hexagonal close-packed
HFS	Heavy Fermion system
IMFP	Inelastic mean free path(s)
IPE	Inverse photoemission
IPES	Inverse photoelectron spectroscopy
LCP	Left circularly polarized
LDOS	Local density of states
LEED	Low-energy electron diffraction
MCDAD	Magnetic circular dichroism in the angular distribution of photoelectrons
MDAD	Magnetic dichroism in the angular distribution of photoelectrons
MDS	Metastable de-excitation spectroscopy
MEED	Medium-energy electron diffraction

ML	Monolayer(s)
MLDAD	Magnetic linear dichroism in the angular distribution of photoelectrons
MOKE	Magneto-optical Kerr effect
NN	Nearest neighbor(s)
PE	Photoemission
PES	Photoelectron spectroscopy
PM	Paramagnetic
QMA	Quadrupole mass analyzer
RCP	Right circularly polarized
RE	Rare earth
RI	Resonance ionization
RKKY	Ruderman–Kittel–Kasuya–Yosida
SEM	Scanning electron microscope/microscopy
SPEELS	Spin polarized electron energy-loss spectroscopy
SPIPE	Spin polarized inverse photoemission
SPLEED	Spin polarized low-energy electron diffraction
SPMDS	Spin polarized metastable de-excitation spectroscopy
SPSTS	Spin polarized scanning tunneling spectroscopy
SPUPS	Spin polarized ultraviolet photoelectron spectroscopy
STM	Scanning tunneling microscope/microscopy
STS	Scanning tunneling spectroscopy
UHV	Ultrahigh vacuum
UMDAD	Unpolarized magnetic dichroism in the angular distribution of photoelectrons
UPS	Ultraviolet photoelectron spectroscopy
VUV	Vacuum ultraviolet
XMCD	X-ray magnetic circular dichroism

Chapter 1
Introduction

The continuing progress in material and surface science opens the possibility to dive into the world of nanostructured systems. A strong stimulation in this fascinating area is given by the computer industry: every device like transistors must get smaller, be more reliable, have a shorter access time and be less energy consuming. In this field of "nanoelectronics" new types of devices are needed leading, e.g., to the development of a single electron transistor. Another important branch of the computer industry consists of the improvement in data storage. The amount of data grows exponentially. Thus, new types of non-volatile memory devices must be constructed mostly basing on a magnetic labeling of the memory elements. One key word in this context is the Giant Magnetoresistance (GMR) found by P. Grünberg and A. Fert. For their discovery both were awarded with the Nobel Prize in Physics in 2007. These considerations already explain the growing interest in the field of "nanomagnetism" which is of course closely related with surface magnetism due to the often dramatically enhanced surface-to-volume ratio of nano-scaled objects.

The conquest of the "nanoworld" was made feasible by new experimental techniques. Several of the highlighting tools are scanning probe methods; a suitable sample is investigated with high lateral resolution by local probes. The first instrument demonstrating the ability to obtain real space images on the atomic scale is the scanning tunneling microscope (STM) [1, 2]. The importance is demonstrated by the Nobel Prize in Physics which was awarded to its inventors Gerd Binnig and Heinrich Rohrer in 1986. This instrument combines the possibility to determine structural and electronic properties on the atomic level with the ability of scanning. Thus, it is the ultimate tool in the field of nanoelectronics. Succeeding to make the STM sensitive to the electron spin polarization opens the door to the world of "nanomagnetism".

The STM allows to probe surface properties with high lateral resolution. A problem concerning local probes is how representative the results obtained on a local scale are, or speaking in other words: Are "typical" images or "typical" spectroscopic information also "typical" on the macroscopic level? The

M. Getzlaff, *Surface Magnetism*, Springer Tracts in Modern Physics, 240,
DOI: 10.1007/978-3-642-14189-8_1, © Springer-Verlag Berlin Heidelberg 2010

complementary problem arises for techniques which average over a macroscopic area. Here it is often unknown how the macroscopic information is composed of the different microscopic contributions. Therefore, only the combination of local probes and macroscopically averaging techniques enables to get a deeper insight into structural, electronic, and magnetic properties. Additionally, one should keep in mind that for the understanding of ferromagnetism on a microscopic level the magnetic phenomena have to be explained in terms of the electronic structure.

To obtain a deeper insight into surface magnetic properties experimental techniques are required which are sensitive to the topmost layers of samples and additionally to the magnetism, i.e. to the electron spin polarization. Some related experimental aspects will be discussed in Chap. 2. First, the focus will be on spin resolving photoelectron spectroscopy and the explanation how to determine the spin polarization of an electron beam. Photoemission with circularly and linearly polarized light enables to obtain information on magnetic properties without spin analysis. These techniques, however, require a "chirality" in the geometrical arrangement which will be presented as well as the sophisticated instruments which allow to obtain tunable radiation with a high degree of circular polarization. The de-excitation of neutral spin polarized noble gas atoms can be used in order to enhance the surface sensitivity due to a tunneling process in the mechanisms belonging to the de-excitation. These processes will be discussed and described how to make this technique sensitive to the electron spin polarization. The physical background and the different modes of operation of a scanning tunneling micro-scope will be described at the end of this part. This instrument was additionally upgraded in order to realize spin polarized vacuum tunneling.

It prevails in today's opinion that understanding of magnetic and electronic behavior requires the knowledge of the structure on a microscopic scale. Due to this important relationship rare earth metal systems will be discussed in Chap. 3 in terms of the dependence of electronic on structural properties. The discussion will start with the pure metals Gd and Tb. The annealing conditions of these films were found to have a strong influence on the magnetic properties. Therefore, one fun-damental aspect will be the determination how the related electronic properties depend on film thickness and morphology. The discussion of binary intermetallics, consisting of rare earth and ferromagnetic $3d$ metals, will follow. This class of materials exhibits outstanding magnetic properties and are additionally of intense technological interest. While the magnetic and magnetostrictive properties of the bulk material in dependence of the fabrication process are well understood very little is known about thin film properties. The experiments presented in the second part of this chapter therefore turn the attention to the influence of the substrate on the growth behavior and the resulting consequences on the crystallographic structure. Further, ternary alloys with a rare earth metal as one of the constituents and with extraordinary behavior, so-called Heavy Fermion Systems (HFS), con-stituting a widely studied class of strongly correlated systems will be discussed. Thermodynamic investigations of these materials result in properties at low tem-perature which are described by a very high effective mass of the electrons. The reason for that is the high correlation of the f electrons with the conduction electrons. While the thermodynamic investigations have more integral character, electron spectroscopic methods can principally offer direct insight into the electronic properties. The most important problem for these measurements is the preparation of a clean and ordered surface. Up to now, these surfaces were generally prepared by scratching with a diamond file or fracturing. These preparation procedures only allow to observe the density of states, but prevent the determination of $E(k)$ dependencies. It will be reported on crystalline thin films of HFS material with Ce as rare earth metal using angle resolving photoemission which allows, with a tunability of photon energy, a direct assignment of the orbital nature of the states which hybridize in the valence band.

The "chemistry" of a surface determines the surface electronic properties. Thus, variations, e.g. due to adsorbate atoms, have a fundamental influence. The knowledge of this phenomenon is therefore of great importance for an unambiguous understanding. Some selected results concerning the influence of adsorbates on structure and electronic behavior of rare earth metal surfaces will be presented in Chap. 4. Hydrogen was found to exhibit unusual adsorption characteristics on Gd. The combination of STM and photoemission allows to image the adsorption process and additionally to determine the electronic behavior. This type of atom can be dissolved in the thin film system which results in plastic deformation. The surface modification enables to determine the structural changes inside the material. Using the distinct surface sensitivity of both techniques—scanning tunneling microscopy and photoelectron spectroscopy—it will be shown, by describing the coadsorption of hydrogen and CO on gadolinium surfaces, that their combination allows a significantly more detailed analysis than only one type of experimental technique.

Basing on the knowledge of the former Chapters a detailed discussion on the surface magnetism of itinerant ferromagnetic materials and magnets with localized moments will be given in Chap. 5. Information on the spin resolved band structure of ferromagnetic materials can *directly* be obtained from spin resolving photoelectron spectroscopy. Using polarized radiation spin integrating photoemission techniques already enable to have access to magnetic properties. An enhancement of the surface sensitivity can be achieved using spin polarized metastable deexcitation spectroscopy. In the first part of this Chapter it will be reported on spin dependent transport and surface magnetic properties of itinerant magnetic substrates, thin Fe(110) and Co(0001) films evaporated on W(110), which were investigated by these electron emission techniques. Further, the behavior of adsorbates will be discussed from the point of view whether they change the properties of the surface and whether they "feel" the magnetism of the underlying substrate. This discussion will be carried out for the example of oxygen which adsorbs dissociatively on the above mentioned surfaces. The spectroscopic capabilities of the scanning tunneling microscope open up the fascinating possibility of correlating the local structural and electronic properties with magnetic ones on the atomic scale. Thus, in the second part of this Chapter spin polarized vacuum tunneling will be demonstrated by measuring the asymmetry of the differential tunneling conductance at bias voltages corresponding to the energetic positions of the two spin contributions of an exchange split surface state. This enables the electronic and magnetic structure information to be clearly separated. By mapping the spatial variation of the asymmetry parameter it is possible to observe the nanomagnetic domain structure of Gd(0001) ultra thin films with a lateral resolution on the nanometer scale.

References

1. G. Binnig, H. Rohrer, C. Gerber, E. Weibel, Phys. Rev. Lett. **49**, 57 (1982)
2. G. Binnig, H. Rohrer, Helv. Phys. Acta **55**, 726 (1982)

Chapter 2
Experimental Aspects

In this chapter the important aspects will be described concerning the experimental techniques which were used to be sensitive to surface magnetic properties. The first one is angle and spin polarized ultraviolet photoelectron spectroscopy (SPUPS) which enables to gather information directly concerning the spin resolved band structure of ferromagnetic materials. In more detail the technique of spin polarized metastable de-excitation spectroscopy (SPMDS) will be discussed with the emphasis on the preparation of a highly spin polarized noble gas atom beam and the possibility to obtain magnetic information in a very surface sensitive way. The third technique is scanning tunneling microscopy (STM) which allows to image the sample "topography" and additionally to determine the electronic and magnetic properties down to the atomic scale. Further, the preparation of atomically clean and well-ordered ferromagnetic surfaces by molecular beam epitaxy of thin film systems is described.

2.1 Photoelectron Spectroscopy

Electron spectroscopy is a very powerful and well established tool to investigate the electronic structure of solids [1]. Angle resolving photoelectron spectroscopy directly enables to yield results on the electronic band structure. The use of tunable photon energies allows a band mapping across the Brillouin zone.

In the following the focus will be on such aspects which are important for the determination of magnetic properties. Being sensitive to the electron spin polarization it is possible to obtain information on the spin dependent and thus magnetic behavior of ferromagnetic materials. The experimental setup is described in Sect. 2.1.1. Without spin analysis valuable information can be gained if polarized radiation is used in combination with a "chiral" arrangement in the experiment. These experimental requirements will be discussed in Sect. 2.1.2. Generating tunable and circularly polarized radiation needs sophisticated instruments. These will be described in Sect. 2.1.3.

M. Getzlaff, *Surface Magnetism*, Springer Tracts in Modern Physics, 240,
DOI: 10.1007/978-3-642-14189-8_2, © Springer-Verlag Berlin Heidelberg 2010

2.1.1 Analysis of Electron Spin Polarization

Angular resolution of the photoelectrons (acceptance cone $\pm 3°$) was made possible by a special home-made electron–optical system with a combination of a 180° and a subsequent 90° deflector. A schematic view of the experimental setup is shown in Fig. 2.1. Owing to their high pass energies they act as quasi-dispersionless elements. Due to this geometrical arrangement, variations to the collection angle were feasible by rotating the whole deflector system. This allowed the subsequent electron spectrometer and spin polarimeter to be kept fixed in space. A cylindrical mirror analyzer with sector field (300 mm slit-to-slit distance) with a high transmission served as the energy dispersing element. The resolution was set to about 300 meV. Magnetic fields were compensated to less than 0.1 μT by Helmholtz coils and additionally by three layers of μ-metal foil inside the vacuum chamber.

Spin analysis was carried out by a Mott-scattering detector with a spherically symmetric acceleration field operated at typically 70 keV without retarding potentials [3]. Surface barrier detectors were used as electron detectors. The figure of merit $S^2 \times I/I_0$ amounts to about 2.4×10^{-4}. The advantages of this type of

Fig. 2.1 Schematic view of the experimental setup for angular and spin resolved photoelectron spectroscopy (from [2], used with permission)

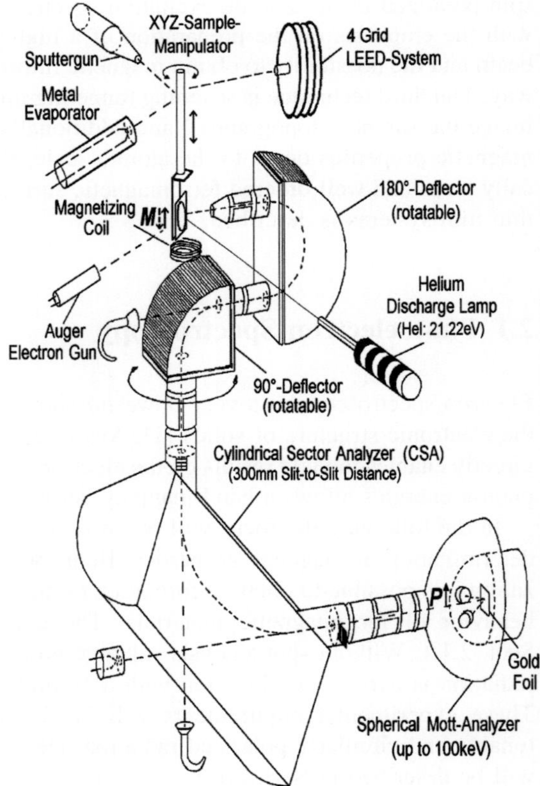

detector are its time-independent asymmetry function, the relatively compact setup, and its low sensitivity to divergent or those electron beams which are shifted off the symmetry–axis. The radii of the inner and outer spheres of the acceleration field were 35 and 70 mm, respectively.

2.1.2 Geometric Arrangement for Dichroism Investigations

The circular dichroism in the angular distribution of photoelectrons (CDAD) [4] is the analogue to the classical circular dichroism in photoabsorption exhibiting different absorption coefficients for left and right circularly polarized light. CDAD in photoemission can only occur, i.e. the change from left to right circularly polarized light results in different photoelectron intensities, if the experimental arrangement has a definite chirality. It is the proper choice of electron collection and photon propagation direction which breaks the symmetry and gives rise to dichroitic effects. A schematic drawing of the geometrical arrangement is shown in the left part of Fig. 2.2. It is obvious that in normal emission ($\theta_e = 0°$) no CDAD effect occurs.

Magnetic circular and linear dichroism in the angular distribution of photoelectrons (MCDAD and MLDAD) are due to the interaction between spin–orbit coupling and exchange splitting. Thus, the intensity distribution of the emitted photoelectrons becomes modified as a function of the macroscopic magnetization direction and the photon spin. The magnetic dichroism is an effect which requires a chirality of the system.

In the case of MCDAD this chirality is induced by the photon propagation direction; the magnetization therefore must have a non-vanishing component along the photon spin. A coplanar arrangement is sufficient for this experimental

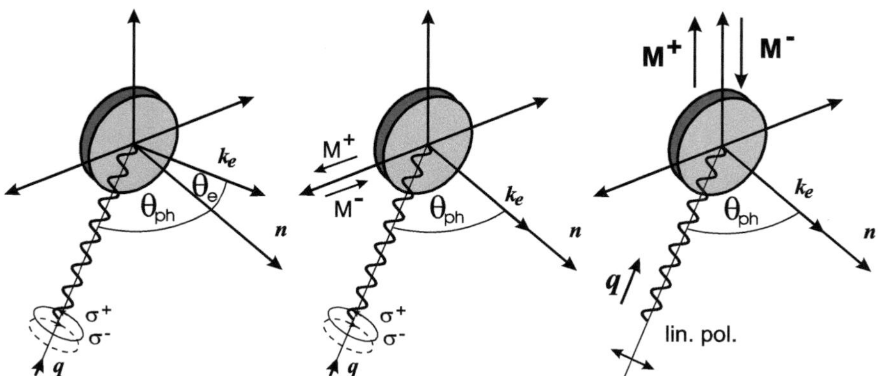

Fig. 2.2 Geometrical arrangements being necessary in photoemission experiments using dichroic effects. *Left* CDAD. *Middle* MCDAD. *Right* MLDAD. q denotes the photon propagation direction, θ_{ph} the angle of the incoming photon beam, k_e the photoelectron momentum, n the surface normal, and $M^{\pm}$ the magnetization direction

technique; this is shown in the middle part of Fig. 2.2. The magnetization was carried out in the plane of the incoming photon beam and the surface normal being the detection direction. Therefore, the MCDAD effect vanishes in normal emission if the magnetization possesses no component along the photon spin.

The MLDAD effect on the other hand requires a chirality induced by the experimental setup itself (for normal emission measurements); i.e. the magnetization must have a component perpendicularly to the plane of photon beam and surface normal. This arrangement is shown in the right part of Fig. 2.2.

The geometrical arrangement indicates the relationship to the magneto-optical Kerr effect (MOKE). MCDAD corresponds to the longitudinal, MLDAD to the transversal MOKE. Additionally, the Faraday effect can be included; it is the analogue to the magnetic circular dichroism in absorption (XMCD).

2.1.3 Circularly Polarized Light

The experimental techniques of CDAD and MCDAD require circularly polarized light in the range of 10 to about 35 eV for investigations of valence band effects and with higher energies in order to determine core level properties. For the low energy region a normal incidence monochromator at BESSY (Berlin) was used [5]. The out-of-plane radiation is elliptically polarized; the degree of circular polarization reaches values up to about 92%. Two apertures as polarization selectors allow to switch from right (RCP) above the plane to left circularly polarized (LCP) light below the plane (see Fig. 2.3). The focussing element consists of a spherical

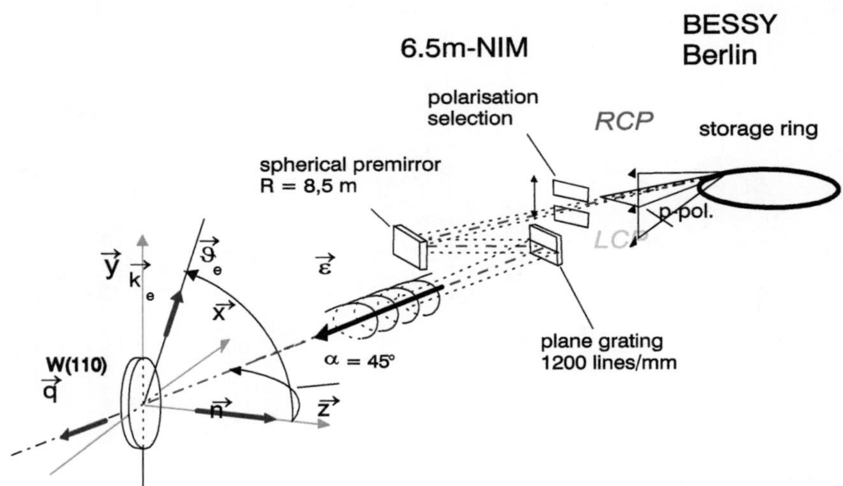

Fig. 2.3 Geometrical arrangement of a normal incidence monochromator. The important directions of the experiment are additionally shown

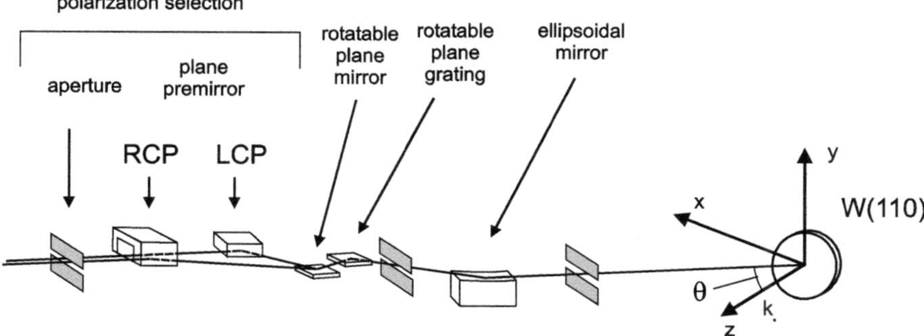

Fig. 2.4 Geometrical arrangement of a plane mirror monochromator for circularly polarized light in the soft X-ray region

premirror. The monochromatization is carried out by a plane grating. Studies during the single bunch mode with ring currents below 100 mA allow to use the linearly polarized light in the plane with the same monochromator. The energy region is limited to about 40 eV due to the normal incidence optics.

For higher photon energies all optical elements must be used in grazing incidence to acquire experiments with a high degree of circular polarization. This is carried out at a plane mirror monochromator at BESSY (see, e.g., [6]). The geometric arrangement is schematically shown in Fig. 2.4. The beamline consists of two separate premirrors for RCP and LCP light. The useful energy region ranges from about 100 to 2,000 eV.

2.2 Metastable De-excitation Spectroscopy

In recent years numerous investigations of clean and adsorbate covered substrates have been carried out by different methods. As most investigations use methods which give information about the behavior in at least a few layers below the surface, there is, in comparison, not so much knowledge about the electronic properties *at the surface*. A *distinct surface sensitivity* can be achieved by electron emission caused by impact of metastable noble gas atoms, a method called metastable de-excitation spectroscopy (MDS) (see, e.g., [7–9]). This technique probes predominantly the outermost atomic layer which will be demonstrated in Sect. 5.1.2 in Chap. 5.

The use of an electron spin polarized metastable $He(2^3S)$ atomic beam is an excellent tool to gain additionally information on *surface magnetic* properties because of the spin selectivity in the de-excitation process. The SPMDS method [10–13] was experimentally pioneered in studies of Ni(110) at Rice University [10]; there, energy analyzing was done by means of a retarding grid yielding the *integrals* of the energy distribution of emitted electrons. The spectrometer used in this investigation allows to determine *directly* the energy distribution of the emitted electrons [14].

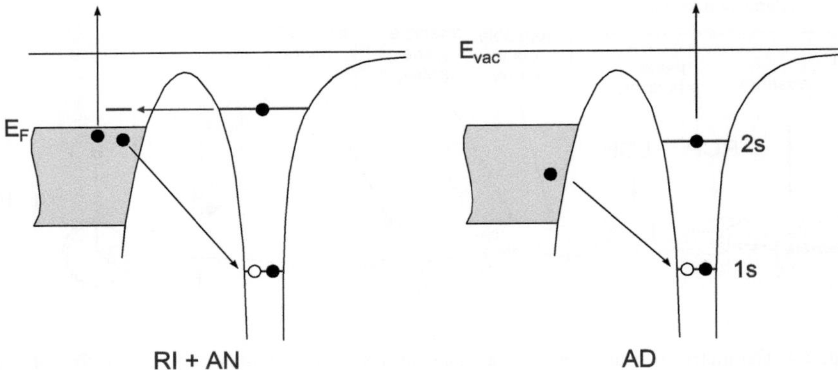

Fig. 2.5 De-excitation mechanisms: resonance ionization RI with a subsequent Auger neutralization AN *left*; Auger de-excitation AD *right*

2.2.1 De-excitation Mechanisms

Depending on the electronic structure of the surface the de-excitation of metastable helium atoms occurs either by resonance ionization (RI) with a subsequent Auger neutralization (AN), or by Auger de-excitation (AD). These mechanisms are schematically shown in Fig. 2.5 and explained below in more detail.

2.2.1.1 Auger Neutralization

If the wave function of the excited electron in the He(2^3S) atom sufficiently overlaps with a degenerate empty level of the solid, then tunneling into this state will occur (RI). The resulting positive ion continues towards the surface where AN takes place in which an electron from the solid tunnels into the $1s$ hole of the helium ion; the energy released is transferred to another electron which may be ejected from the solid. The kinetic energy of the escaping electron depends on the effective binding energy of the $1s$ hole. This energy decreases as the ion approaches the surface due to the image potential. Thus, the closer to the surface this process occurs, the smaller the energy of the emitted electron. Its kinetic energy is given by

$$\begin{aligned}
E_{\text{kin}} &= E_{1s}^{\text{eff}} - (\Phi + \bar{E}_B + \varepsilon) - (\Phi + \bar{E}_B - \varepsilon) \\
&= E_{1s}^{\text{eff}} - 2(\bar{E}_B + \Phi)
\end{aligned} \tag{2.1}$$

with Φ being the work function and introducing the averaged binding energy $\bar{E}_B$ of both electrons being involved in the process (see Fig. 2.6). E_{1s}^{eff} represents the effective energy of the $1s$ state being different to the energy of the *free* atom due to interaction with the surface.

Every kinetic energy E_{kin} therefore corresponds to two electrons with binding energies being symmetrically to $\bar{E}_B$. Thus, the energy distribution cannot *directly*

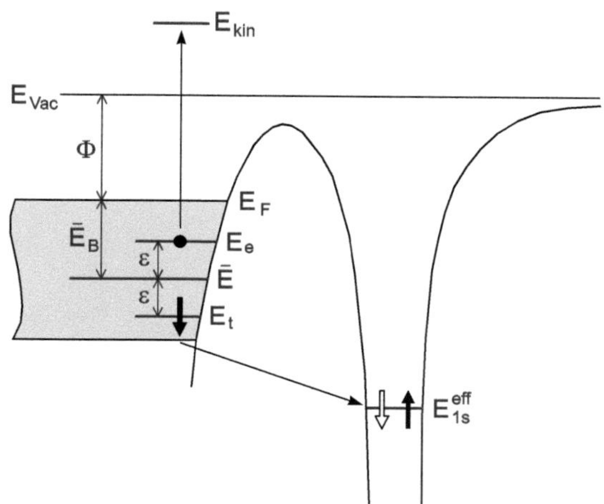

Fig. 2.6 Energetic scheme of Auger neutralization AN

be interpreted as the density of states. An exception occurs at the onset for high kinetic energies. E_{kin} exhibits its maximum if $\bar{E}_B = 0$, i.e. both involved electrons are energetically localized at the Fermi level:

$$E_{kin}^{max} = E_{1s}^{eff} - 2\Phi \qquad (2.2)$$

Additionally, the comparison with photoemission spectra allows to determine the effective energy of the $1s$ state E_{1s}^{eff} to be:

$$E_{1s}^{eff} = h\nu - \Delta E_{kin}^{max} + \Phi \qquad (2.3)$$

with $\Delta E_{kin}^{max} = E_{kin,UPS}^{max} - E_{kin,AN}^{max}$. This will be demonstrated in Sect. 5.1.2.

2.2.1.2 Auger De-excitation

If the metastable atoms get close to the surface without RI taking place, then AD can occur. This is the dominant de-excitation process if RI is suppressed, which occurs if the level of the excited helium atom lies below the Fermi level ($I_{eff} > \Phi$ with I_{eff} being the ionization potential, see Fig. 2.7) or if there is an insufficient overlap with empty states due to an adsorbate layer. In this case the $1s$ hole is filled by an electron from the solid or the adsorbate layer with a simultaneous ejection of the excited $2s$ atomic electron.

2.2.2 Sensitivity to Spin Polarization

The occurrence of a difference in the intensities of the ejected electrons with different spin directions by interacting of a *spin polarized* metastable He(2^3S)

Fig. 2.7 Energetic scheme
of Auger de-excitation AD

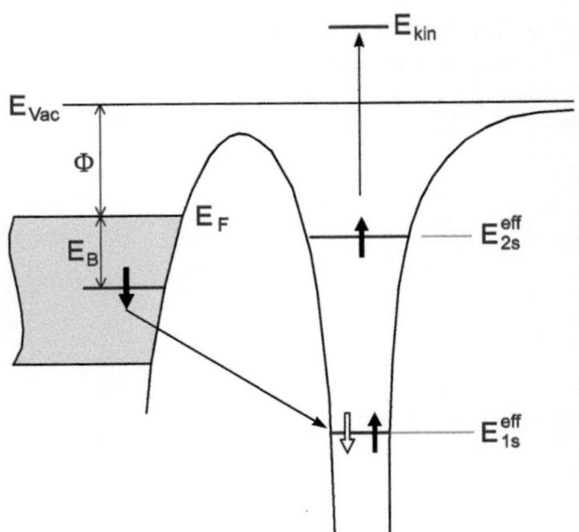

beam with a *ferromagnetic* surface is illustrated in Fig. 2.8, exemplarily for the AD process. If AN takes place, the de-excitation process is, in principle, similar but more complicated due to the fact that *two* electrons of the valence band are involved. The density of states (DOS) can then be obtained by a deconvolution.

In the depicted case in Fig. 2.7 *only* a *minority* electron can fill the $1s$ hole, when exciting with $He(2^3S)\uparrow$ atoms, leading to an intensity $I^\uparrow$ of the escaping $2s$ electrons. By exciting with $He(2^3S)\downarrow$ atoms, only *majority* electrons can tunnel to the helium atom leading to an electron intensity $I^\downarrow$. For *ferromagnetic materials* the spin resolved densities of states are *different* for majority and minority electrons (schematically shown on the left hand side of Fig. 2.8). Thus, the asymmetry A, normalized to the polarization of the incident atomic beam with polarization P_A, corresponds to the difference in the DOS for the majority and minority electrons:

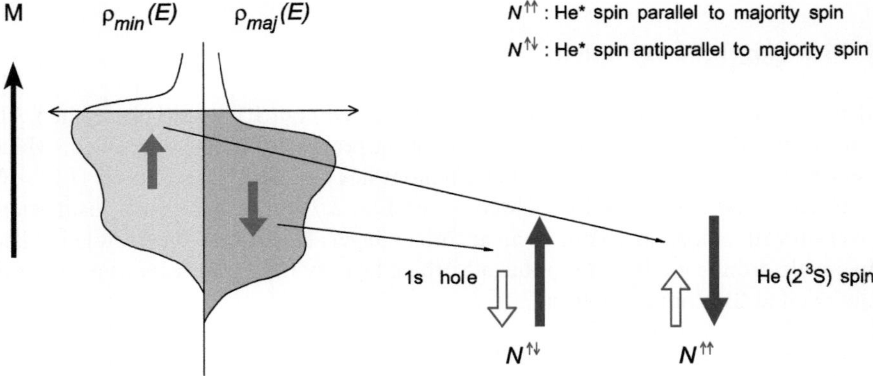

Fig. 2.8 Auger de-excitation process (for details see text) for a spin polarized metastable He beam interacting with a ferromagnetic surface ($I_{eff} > \Phi$). Due to the spin selectivity of this process, differences in the spin resolved density of states in the outermost layer (schematically shown on the *left* hand side) lead to different intensities of ejected electrons with majority ($\uparrow$) or minority ($\downarrow$) spin

$$A = \frac{1}{P_A} \cdot \frac{I^\uparrow - I^\downarrow}{I^\uparrow + I^\downarrow} \propto \frac{D(E)^\downarrow - D(E)^\uparrow}{D(E)^\downarrow + D(E)^\uparrow}. \qquad (2.4)$$

Due to the surface sensitivity of MDS and the spin selectivity of the de-excitation process, properties of the *spin resolved* DOS of the outermost layer can therefore be obtained.

2.2.3 *Preparation of Spin Polarized Metastable Atoms*

2.2.3.1 Experimental Setup

The MDS experiment is performed in an ultrahigh vacuum (UHV) chamber with a base pressure in the upper 10^{-11} mbar range. The chamber is equipped with a combined low-energy electron diffraction (LEED)—Auger system for surface characterization and a quadrupole mass analyzer (QMA). The source of spin polarized metastable He atoms is described below. The angle of the incoming atomic beam is $\theta_i = 30°$ with respect to the surface normal. Determining the degree of polarization is carried out by means of an analyzing sextupole magnet in conjunction with a spin flipper. Photoelectron spectra are obtained with unpolarized vacuum ultraviolet (VUV) radiation from a discharge lamp (He I: $hv = 21.22$ eV); the angle of incidence is $60°$. All spectra are taken in normal emission. The electron–optical entrance system provides an angular resolution of the emitted electrons (acceptance cone $\pm 5°$). A $150°$ spherical spectrometer with a radius of 100 mm serves as the energy dispersing element. The resolution is set to about 150 meV. Detection of the energy resolved electrons is achieved by a channel-electron multiplier (CEM). Magnetic fields are compensated by three pairs of Helmholtz coils.

2.2.3.2 Discharge Source

A helium *discharge source* produces a beam which contains ground-state He(1^1S) atoms, metastable He(2^1S) and He(2^3S) atoms (He*) as well as long-lived Rydberg atoms. The radiative lifetime of the He(2^3S) atoms of about 10^4 s [15] is "infinitely" long compared with the millisecond flight time between source and detector.

2.2.3.3 Sextupole Magnet

A subsequent *sextupole magnet* (the magnetic arrangement is given in Fig. 2.9) acts as a *polarizer* by preferentially transmitting the atoms with a magnetic

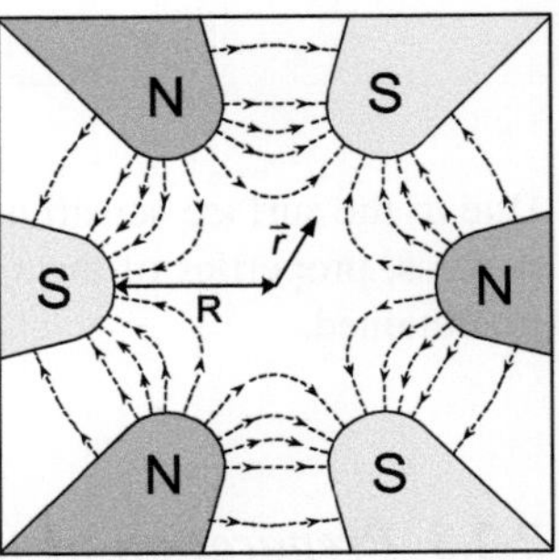

Fig. 2.9 Magnetic fields of the sextupole magnet

quantum number $m_s = +1$ and strongly suppressing the ones with $m_s = -1$. For atoms with a field–strength-independent magnetic moment, such as He(2^3S), the inhomogeneous magnetic field in the pole gap of a sextupole magnet provides a force which is proportional to the atoms' radial distance from the axis and is directed towards it if $m_s > 0$, away from the axis if $m_s < 0$, and no force is acting if $m_s = 0$. For atoms with $m_s > 0$, this magnet acts like a positive lens. This lens is free of aberrations except for the chromatic one which restricts the focussing for thermal atomic beams to a portion of their velocity distribution. Without a central stop at the magnet exit, the atoms with $m_s = 0$, including all the ground-state and the metastable singlet atoms, are transmitted if their straight-line trajectories can clear the magnet. With a central stop in place, ideally only the He(2^3S) atoms in Zeeman state $m_s = +1$ are transmitted.

2.2.3.4 Spin Flipper

The measurements were carried out with antiparallel and parallel, respectively, spin orientations of the atomic beam electrons **P** and of the electrons of the ferromagnetic surface. Reversing the polarization of the He atom beam by reversing the magnetic guiding field is not acceptable because the field change can result in a large systematic error by affecting the intensity and the position of the electron beam. The use of a *spin flipper* [16] provides a method for reversing the polarization of the incoming atomic beam while keeping the magnetic guiding field constant. Such a reversal corresponds to a Zeeman transition from $m_s = +1$ to $m_s = -1$ [16]. The principle of spin motion is schematically shown in Fig. 2.10. The axial field strength inside the spin flipper and its actions in the diabatic and adiabatic mode of operation are described in detail in [17]. The efficiency of the spin flipper is determined to be very close to 100%, that is, one obtains equal beam polarization values for the reversed and the not-reversed case.

Figure 2.11 shows the profiles of the polarized beam, taken by a Stern–Gerlach dipole magnet, for the adiabatic (a) and (c) and the diabatic (b) and (d) setting of the spin flipper, each taken without (a) and (b) and with the central stop (c) and (d) at the polarizing sextupole exit. Evaluation of the profiles of Fig. 2.11c and 2.11d

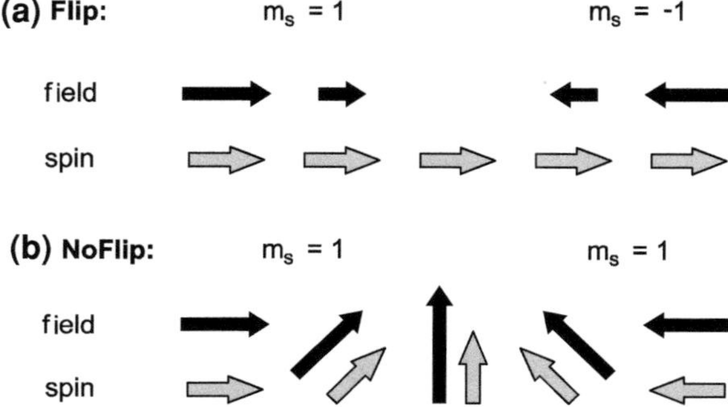

Fig. 2.10 a Spin motion in the diabatic case (schematically). **b** Spin motion in the adiabatic case with an additional transverse field of 100 μT applied in the center of the spin flipper

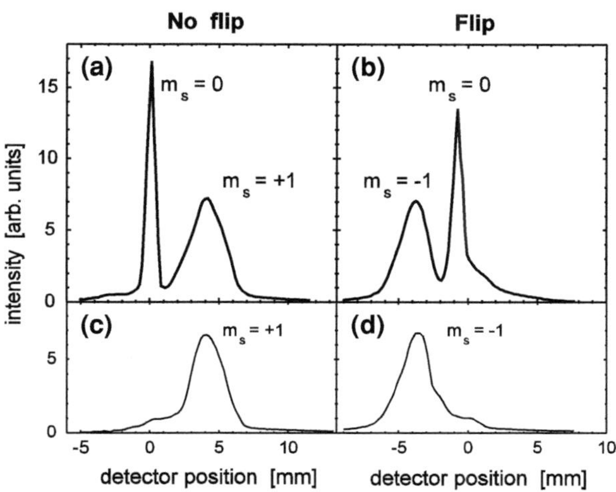

Fig. 2.11 Beam profiles of the metastable He beam: **a** without central stop, adiabatic mode of the spin flipper; **b** without central stop, diabatic mode; **c** with central stop, adiabatic mode; **d** with central stop, diabatic mode

with respect to their composition of atoms with $m_s = +1$, $m_s = 0$, and $m_s = -1$ components yields the He(2^3S) beam polarization of $P_A = (0.90 \pm 0.02)$. The flux near the target region amounts to about 5×10^{11} s^{-1}, the density to about 1×10^7 cm^{-3} [17].

2.2.3.5 Magnetic Guiding Field

Additionally, a magnetic guiding field is necessary to prevent depolarization. Between the components along the beamline the guiding field is provided by a longitudinal field of about 100 μT. Within the sextupole magnet the ferromagnetic poles provide strong inhomogeneous fields orthogonal to the beam axis overriding the weak guiding field. The spin flipper is surrounded by a magnetic shield which keeps the guiding field outside and allows to establish conditions for either adiabatic or diabatic transitions.

2.3 Scanning Tunneling Microscopy

In a scanning tunneling microscope a conducting sample and a sharp metallic tip are brought within a distance of a few Ångström resulting in a significant overlap of the electronic wave functions. With an applied bias voltage a tunneling current can flow from the occupied electronic states of one electrode into the empty states of the other one. Using a piezoelectric drive system for the tip and a feedback loop a map of the "surface topography" can be obtained. The high spatial resolution results from the exponential dependence of the tunneling current on the tip-sample distance. Thus, an STM can provide real space images of surfaces down to the atomic scale.

2.3.1 Vacuum Tunneling

From the view of classical mechanics a microscopic particle, e.g. an electron, cannot traverse a potential barrier which is higher than its kinetic energy. Using quantum mechanics, however, we know that this is possible due to the wave–particle dualism. The probability to find the electron on the other side of the barrier is not equal to zero if it impinges the barrier. The transmission coefficient T representing the ratio of transmitted and incident current is given by

$$T = \frac{1}{1 + (k^2 + \kappa^2)^2 / 4k^2\kappa^2 \sinh^2(\kappa s)} \tag{2.5}$$

with E the energy of the incoming electron and V_0 the energetic height of a rectangular potential barrier exhibiting the width s. The decay rate κ amounts to

$$\kappa = \frac{\sqrt{(2m(V_0 - E))}}{\hbar} \tag{2.6}$$

Assuming $\kappa s \gg 1$ being fulfilled for usual tunnel voltages and distances Eq. 2.5 can approximately be written as

$$T \approx \frac{16k^2\kappa^2}{(k^2 + \kappa^2)^2} e^{-2\kappa s} \tag{2.7}$$

The exponential dependence of the transmission coefficient and thus of the tunneling current on the distance is the reason for the high lateral resolution of an STM. This dependence results in a tunneling current which is dominated by the very first atoms at the probe tip. A change of the barrier width of only 2 Å (the typical layer distance of metals) results in a decrease of the tunneling current to about 1%. Thus, combining a high mechanical stability with a sufficient positioning accuracy sample surfaces can be imaged with atomic resolution.

This description principally allows to understand the high lateral resolution of an STM but it fails, for example, to explain a chemical contrast. An improved model was given by Tersoff and Hamann [18, 19]. Using the time-dependent perturbation theory of Bardeen [20], a Fourier development for the wave functions of the sample surface and assuming an s-like orbital as probe tip neglecting contributions from tip wave functions with angular dependence (orbital quantum number $\ell \neq 0$), the tunneling current can be written, for small applied bias voltages and low temperatures, as

$$I \propto U \cdot \rho_\mathrm{s}(E_\mathrm{F}) \cdot e^{2\kappa R} \cdot \sum_v |\Psi_v(\mathbf{r}_0)|^2 \cdot \delta(E_v - E_\mathrm{F}) \qquad (2.8)$$

with $\kappa = \sqrt{2m\Phi}/\hbar$ being the decay rate and Φ the effective local barrier height. A close look to this expression results in the following characteristics of the tunneling contact and the topographic imaging:

- The tunneling contact exhibits an Ohmic behavior, i.e. I is proportional to U.
- The tunneling current is proportional to the density of states at the Fermi level of the probe tip $\rho_\mathrm{T}\,(E_\mathrm{F})$.
- The term $\sum_v |\Psi_v(\mathbf{r}_0)|^2 \cdot \delta(E_v - E_\mathrm{F})$ describes the local density of states of the sample below the tip apex which is therefore proportional to the tunneling current I. The exponential dependence on the distance (see Eq. 2.7) is implicitly included in the wave function of the sample Ψ_v being proportional to $\exp(-\kappa z)$ and thus decreasing exponentially into the vacuum.

This simple consideration is no longer valid for higher applied bias voltages or tip functions with angular momentum. A generalization was carried out by Chen [21]. As already mentioned (cf. Eq. 2.8), the tunneling current is proportional to the density of states of the sample below the tip apex. The applied bias voltage determines which states of the sample contribute to the tunneling current due to the energetically and bias voltage dependent transmission coefficient [22]:

$$I \propto \int_0^{eU} \rho_\mathrm{s}(\pm eU \mp E)\rho_\mathrm{p}(E)T(E, eU)\,dE \qquad (2.9)$$

with

$$T(E, eU) = \exp\left\{ -2(d + R)\left[\frac{2m}{\hbar^2}\left(\frac{\Phi_\mathrm{s} + \Phi_\mathrm{p}}{2} + \frac{eU}{2} - E\right)\right]^{1/2} \right\} \qquad (2.10)$$

The states directly at the Fermi level therefore mostly contribute to the tunneling current.

2.3.2 Modes of Operation

In the following the focus will be on different modes of operation and the explanation which information can be obtained.

2.3.2.1 Topographic Mode

The mostly used mode of operation is the *constant current mode*. In this mode the tip is stabilized using a feedback loop system in such a way that the tunneling current between tip and sample is held constant. This is carried out using a piezoelectric driver system which changes the distance. The voltage which is applied at the piezo driver is recorded as a function of the position of the tip resulting in a topographic image. If the sensitivity of the driver is calibrated the topographic height of the sample can directly be obtained.

The principle seems to be rather simple but the obtained image of the sample surface is not solely the arrangement of atoms on the surface. The contour map rather reflects the map of constant current being determined by the local density of states. For example, oxygen atoms *on* the surface are imaged as *depressions*. A further example, hydrogen on gadolinium surfaces, is discussed in Sect. 4.1 in Chap. 4.

2.3.2.2 Spectroscopic Mode

For low bias voltages the tunneling current is linearly proportional to the applied voltage. For higher ones the bias dependence of the tunneling current does not exhibit this Ohmic behavior.

This leads to further information which can be obtained by using the spectroscopic mode of operation. In this mode the tunneling current is determined as a function of the applied bias voltage thus additionally allowing to determine the electronic structure of the sample surface down to the atomic scale.

For zero applied bias the Fermi levels of tip and sample are equal. Applying a bias voltage U to sample or tip results in a shift of the energy levels by an amount of $|eU|$ depending on the polarity. For positive sample bias the net tunneling current arises from electrons which tunnel from the occupied tip states into empty sample states whereas for negative sample bias electrons tunnel from occupied sample states into empty tip states. Varying the value of applied bias voltage therefore allows to identify the electronic states which contribute to the tunneling current. The current increases significantly if the applied bias voltage enables a tunneling into an empty state. Thus, in a first approximation the derivative of the tunneling current with respect to the applied bias voltage is a measure of the local density of states. The transmission coefficient increases monotonically with the applied bias voltage. Thus, it results in a smoothly varying background which the features are superimposed on being due to electronic states.

Assuming a structureless density of states for the tip the local density of states for a sample is given by the first derivative of the tunneling current with respect to the applied bias voltage:

$$\frac{dI}{dU} \propto \rho_p(E)T(E,eU)dE \tag{2.11}$$

This so-called differential conductance dI/dU allows to determine the local density of states of the sample with high lateral resolution. The differential conductance can directly be determined using Lock-In technique. A periodic voltage is added to the bias voltage and the modulation of the tunneling current with the same frequency is measured with the feedback loop being disabled. The signal-to-noise ratio using Lock-In technique is enhanced by a factor of about 100 compared to a numerical derivation of the tunneling current with respect to the bias voltage as a function of the bias voltage. The influence of the transmission coefficient must be taken into consideration for the interpretation of the data.

The distinct advantage of scanning tunneling spectroscopy (STS), compared to conventional experimental techniques like photoelectron spectroscopy and inverse photoemission, is the high lateral resolution. Additionally, electronic states can be investigated in a *single* measurement on *both* sides of the Fermi level which cannot be carried out in photoemission techniques; photoemission allows to determine occupied states only, inverse photoemission empty states only.

2.4 Sample Preparation

A W(110) crystal served as substrate in all investigations. It was cleaned by heating in oxygen and flashing up to 2,600 K. The magnetic films were evaporated by means of a small electron-beam evaporator. The principle arrangement is schematically shown in Fig. 2.12. An integrated ionization-gauge-like flux monitor facilitated a reproducible growth rate; typically it was set to one layer per minute. By means of a shutter it is possible to start the evaporation process when reaching the regular working conditions. During evaporation not only neutral atoms but also metal ions are produced which are a direct measure of the quantity

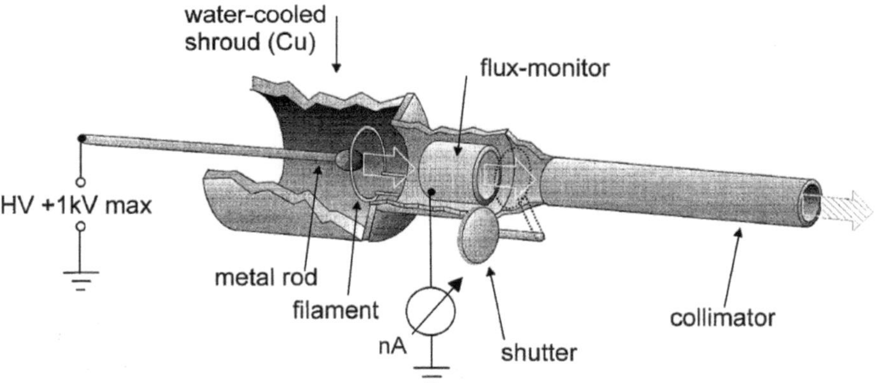

Fig. 2.12 Principle arrangement of the metal evaporator for the epitaxial growth of thin metallic films. The current on the collector, due to metal ions during evaporation, facilitates a constant and reproducible flux with a deviation below 5% (from [2], used with permission)

of evaporated material. The collector is electrically insulated in order to detect the metal ions as current (typically 100 nA).

Iron and cobalt were evaporated using thin rods (typically 2 mm diameter), gadolinium and terbium from a tungsten crucible. During evaporation the pressure stayed in the low 10^{-10} mbar range.

Due to the in-plane surface anisotropy [23] the easy magnetization axis in thin iron films is the in-plane [$1\bar{1}0$] direction being the hard magnetization axis of bulk Fe, switching at a critical thickness to the one of bulk crystals, i.e. the [001] direction. For cobalt films the anisotropy causes the easy magnetization axis to lie in-plane in contrast to bulk-hcp Co with its easy axis perpendicular to the basal plane. Thick rare earth metal films exhibit an easy axis within the surface plane. The shape anisotropy may also change the easy magnetization axis.

For the spin resolving photoemission and spin polarized metastable de-excitation spectroscopy measurements the films were magnetized by a current pulse through a coil close to the sample along the [$1\bar{1}0$] direction of the tungsten substrate.

References

1. S. Hüfner, *Photoelectron Spectroscopy*, 3rd edn. (Springer, Berlin, 2003)
2. M. Getzlaff, J. Bansmann, J. Braun, G. Schönhense, Z. Phys. B **104**, 11 (1997)
3. M. Getzlaff, B. Heidemann, J. Bansmann, C. Westphal, G. Schönhense, Rev. Sci. Instr. **69**, 3913 (1998)
4. C. Westphal, J. Bansmann, M. Getzlaff, G. Schönhense, Phys. Rev. Lett. **63**, 151 (1989)
5. F. Schäfers, W. Peatman, A. Eyers, C. Heckenkamp, G. Schönhense, U. Heinzmann, Rev. Sci. Instr. **57**, 1032 (1986)
6. J. Bansmann, C. Ostertag, G. Schönhense, F. Fegel, C. Westphal, M. Getzlaff, F. Schäfers, H. Petersen, Phys. Rev. B **46**, 13496 (1992)
7. H. Conrad, G. Ertl, J. Küppers, W. Sesselmann, H. Haberland, Surf. Sci. **100**, L461 (1980)
8. B. Woratschek, W. Sesselman, J. Küppers, G. Ertl, H. Haberland, Phys. Rev. Lett. **55**, 611 (1985)
9. W. Sesselmann, B. Woratschek, J. Küppers, G. Ertl, H. Haberland, Phys. Rev. B **35**, 1547 (1987)
10. M. Onellion, M. Hart, F. Dunning, G. Walters, Phys. Rev. Lett. **52**, 380 (1984)
11. M. Hammond, F. Dunning, G. Walters, G. Prinz, Phys. Rev. B **45**, 3674 (1992)
12. H. Hagstrum, Phys. Rev. **96**, 336 (1954)
13. D. Penn, P. Apell, Phys. Rev. B **41**, 3303 (1990)
14. M. Getzlaff, D. Egert, H. Steidl, G. Baum, W. Raith, Z. Phys. D **30**, 245 (1994)
15. C. Drake, in *Atomic Physics III*, ed. by S. Smith, G. Walters (Plenum-Press, New York, 1977), p. 269
16. W. Schröder, G. Baum, J. Phys. E **16**, 52 (1983)
17. G. Baum, W. Raith, H. Steidl, Z. Phys. D **10**, 171 (1988)
18. J. Tersoff, D. Hamann, Phys. Rev. Lett. **50**, 1998 (1983)
19. J. Tersoff, D. Hamann, Phys. Rev. B **31**, 805 (1985)
20. J. Bardeen, Phys. Rev. Lett. **6**, 57 (1961)
21. C. Chen, Phys. Rev. Lett. **65**, 448 (1990)
22. R. Wiesendanger, *Scanning Probe Microscopy and Spectroscopy* (Cambridge University Press, Cambridge, 1967)
23. U. Gradmann, J. Korecki, G. Waller, Appl. Phys. A **39**, 101

Chapter 3
Structural and Electronic Properties
of Rare Earth Metal Systems

Thin film and surface studies of Gd as a prototypical ferromagnetic $4f$ rare earth metal have attracted considerable interest because of its interesting surface magnetic properties. For instance, an enhancement of the Curie temperature of the surface compared to the bulk ($T_{CB} = 293$ K) was observed based on spin polarized low-energy electron diffraction (SPLEED) [1], magneto-optical Kerr effect (MOKE) [2], and spin polarized secondary electron emission spectroscopy studies [3]. The electronic structure of Gd films was investigated by angle resolving photoemission (PE) and inverse photoemission (IPE) experiments [4], as well as by spin resolving photoelectron spectroscopy [5].

A significant change in the $5d$ state and, in addition, of the core levels with decreasing film thickness down to approximately 1.5 monolayers (ML) was revealed [6], as was the existence of a surface state at about 0.3 eV above the Fermi level E_F. Furthermore, a non-vanishing exchange splitting even at room temperature of the unoccupied $5d$ state could be confirmed [7] in agreement with the observation of an enhanced surface magnetic order. The annealing conditions of Gd films, deposited at room temperature, were found to have a strong influence on the magnetic properties, as shown by MOKE [8, 9] and ac magnetic susceptibility measurements on films exhibiting a thickness between 5 and 11 ML [8].

This indicates the *importance of the relationship between topographical and magnetic structure* of the films. Therefore, the first part of this chapter concentrates on:

- How does Gd grow in the submonolayer regime?
- What is the influence of different preparation procedures on the morphology of thick Gd films on W(110)?
- Is it possible to observe the surface state by means of scanning tunneling spectroscopy (STS)?
- Do the electronic properties depend on film thickness and morphology?

M. Getzlaff, *Surface Magnetism*, Springer Tracts in Modern Physics, 240,
DOI: 10.1007/978-3-642-14189-8_3, © Springer-Verlag Berlin Heidelberg 2010

After the description of pure Gd the focus will be on binary alloys of $4f$ rare earth metals with $3d$ transition metals because they exhibit outstanding magnetic properties and are additionally of intense technological interest. The Curie temperature of alloys of Gd or Tb with Fe as well as their coercivity can be tuned over a wide range by changing the mixing ratio [10]. This behavior, in combination with a strong magneto-optical effect, makes them of particular interest for magneto-optical storage. Furthermore crystalline alloys, especially $TbFe_2$, exhibit a strong magnetocrystalline anisotropy resulting in an easy magnetization direction perpendicular to the surface plane in thin film systems [11]. On one hand films with perpendicular magnetization direction allow enhanced writing densities as well as an increased signal-to-noise ratio on the other hand. Another aspect of technological interest are magnetostrictive properties of $3d/4f$ alloys like $TbFe_2$, $DyFe_2$ and $Dy_{0.7}Tb_{0.3}Fe_2$ (known as Terfenol-D). For such applications, the preparation of single crystalline thin films is of importance because a well defined orientation of the crystallographic directions is needed to make use of the magnetostrictive effect in small devices.

While the magnetic and magnetostrictive properties of the bulk material in dependence of the fabrication process, especially for Terfenol-D, are well understood very little is known about thin film properties. Until now only some attempts have been made to grow very thin crystalline films of $3d/4f$-alloys. $TbFe_2$ [12] and YCo_2 [13] have been grown on sapphire with Mo(110) and W(110) buffer layers, respectively. In both cases twin formation occurred. Oderno et al. [14] succeeded in preparing $DyFe_2$, $TbFe_2$ and $Dy_{0.7}Tb_{0.3}Fe_2$ epitaxial thin films on sapphire with a Nb(110) buffer layer. Huth et al. [15] presented results on the preparation of well-ordered epitaxial films of $TbFe_2$ on sapphire substrates with a Mo(110) buffer layer. As far as the pure elements Fe, Gd, and Tb are concerned most of the experiments on the magnetic thin film properties have been carried out for films grown on W(110). The growth of these films from the submonolayer coverage range to thick films is well understood and forms the basis for understanding the magnetic properties.

The experiments presented in the second part of this chapter therefore turn the attention to the following questions:

- What does the initial state of $3d/4f$ alloy formation on W(110) look like?
- Is it possible to grow epitaxial films of $GdFe_2$ on W(110) without additional buffer layers?
- What is the influence of the substrate on the growth behavior and what consequences on the crystallographic structure will result from this?
- Which crystallographic orientation will the film grow in?

After the description of binary alloys the focus will be on ternary alloys with a rare earth metal as one constituent which exhibit extraordinary properties. In the last decades the so-called Heavy Fermion Systems (HFS) [16–18] attracted a growing interest in solid state physics constituting a widely studied class of strongly correlated systems. These systems are binary or ternary compounds with lanthanides or actinides. Thermodynamic investigations of these materials result in properties at low temperature which are described by a very high effective mass of the electrons. The reason of this feature is due to the high correlation of the f electrons with the conduction electrons due to hybridization. While the thermodynamic investigations have more integral character, electron spectroscopic methods can principally offer a direct insight into the electronic properties.

The most important problem for these measurements is the preparation of a clean and ordered surface. Up to now, these surfaces were generally prepared by scratching with a diamond file or fracturing. Other techniques as sputtering by noble gas atoms roughens the surface to an unacceptable degree. These preparation procedures only allow to observe the density of states but prevent the determination of $E(k)$ dependencies.

Especially angle resolved photoemission is fundamental for the determination of electronic properties. In conjunction with a tunability of the photon energy a direct assignment of the orbital nature of the states which hybridize in the valence band can be obtained [19, 20].

In the third part of this chapter it will be reported on crystalline thin films of HFS material with Ce as rare earth metal on a single crystalline surface. This preparation technique enables to prepare *atomically clean* and *well-ordered* surfaces. As ternary intermetallics with rare earth elements $CePd_2Si_2$ and $CeNi_2Ge_2$ were chosen. The electronic behavior of these high-quality films were determined using the technique of resonant photoemission. The giant resonance in the Ce $4f$ emission at 122 eV facilitated the determination of $4f$ derived features in the photoelectron spectra [21] as well as to distinguish between the different hybridization characters of the bands being involved in the bonding of the intermetallics. Electron energy-loss spectroscopy (EELS) allowing dipole-forbidden f–f excitation was used to investigate the binding energies of the Ce $4f$ states due to the creation of a neutral excited state and thus avoiding a final ionic hole.

3.1 Gadolinium as Prototype of a Rare Earth Metal

The extraordinary surface magnetic properties of heavy rare earth metals have been the subject of growing interest during the last decades due to controversially discussed results. Gadolinium can be considered as the prototype ferromagnetic material with localized magnetic moments. Its Curie temperature is the highest one for rare earth metals. The half filled $4f$ shell causes the high magnetic moment of $7\mu_B$.

The discussion will start with the behavior of Gd as a thin film system on W(110) and be followed with that of thicker films. The conclusions drawn will build the basis for understanding the thickness and morphology dependent electronic structure of Gd(0001) which is described subsequently.

3.1.1 Gd on W(110) at Submonolayer Coverage: Structural and Local Electronic Properties

Although Gd(0001) films on W(110) have intensively been studied in the past for the coverage regime of one up to several ML, very little is known about the submonolayer coverage regime. The morphology of the films was first studied by low-energy electron diffraction (LEED) and Auger electron spectroscopy (AES) [22, 23]. These techniques are known to average over at least several tenths of a square millimeter. A more detailed view on the film morphology can be achieved by scanning tunneling microscopy (STM) which yields a real space image of the topography on the relevant scale of several 10 nm up to about 100 nm. An STM study on the growth of Gd on W(110) in the coverage range of one ML up to about 20 ML with the focus on 11 ML was published in [24]. In this first part it will be reported on an STM study on the growth of Gd/W(110) at submonolayer coverage [25]. In favor of comparability the same annealing temperatures as Tober et al. [24] (i.e. 530 and 710 K) were chosen.

Figure 3.1 shows a topographic STM image of a 60 nm $\times$ 60 nm area of a nominally 0.5 ML Gd film on W(110) annealed at 710 K for 10 min. Five different one-dimensional surface structures (labeled a–e) appear in the STM image which correspond to (a) (8 $\times$ 2), (b) (7 $\times$ 2), (c) (6 $\times$ 2), (d) (5 $\times$ 2), and (e) c(5 $\times$ 3) superstructures of the Gd thin film. The superstructures (a–d) result in $(n \times 2)$ LEED pattern with $n = 8, 7, 6, 5$.

Figure 3.2 shows line sections along the [001] direction, i.e. perpendicular to the chains, which were measured at negative bias voltage ($U = -0.8$ V). Obviously, the apparent corrugation decreases with decreasing n. At this particular bias voltage it is up to 1.2 Å for the (10 $\times$ 2) structure but it did not exceed 0.1 Å for the (5 $\times$ 2) structure. However, the measured corrugation of $(n \times 2)$ structures with $n \geq 7$ strongly depends on the applied bias voltage as shown in Fig. 3.2b

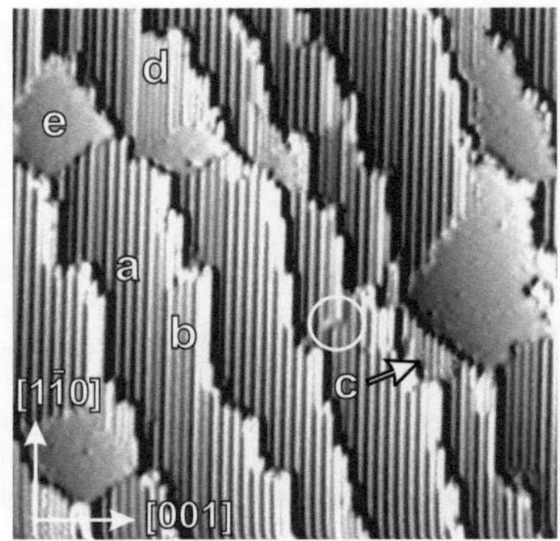

Fig. 3.1 Constant-current STM topograph (scan range: 60 nm $\times$ 60 nm) of 0.5 ML Gd on W(110) annealed at 710 K. Different superstructures are observed at different locations: **a** (8 $\times$ 2), **b** (7 $\times$ 2), **c** (6 $\times$ 2), **d** (5 $\times$ 2), and **e** c(5 $\times$ 3). The *circle* marks a dislocation. Tunneling current $I = 1$ nA, sample bias voltage $U = -0.8$ V. Reprinted with permission from [26]. Copyright (1997) by the American Physical Society

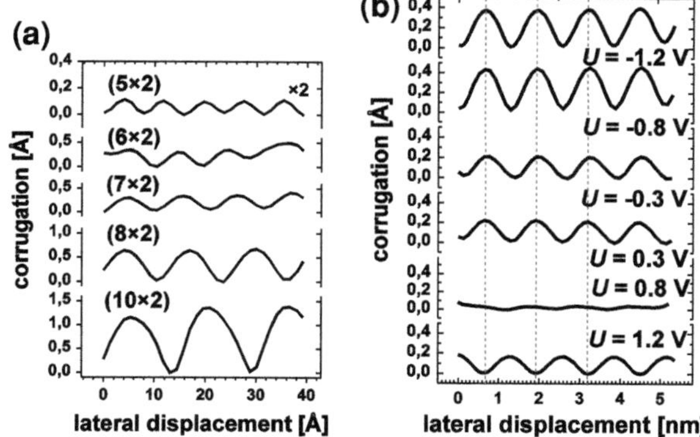

Fig. 3.2 **a** Line sections drawn along the [001] direction, i.e. perpendicular to the chains. The data for all $(n \times 2)$ superstructures except for $n = 10$ ($I = 1$ nA, $U = -0.8$ V) were extracted from the same STM image ($I = 1$ nA, $U = -0.8$ V). **b** Bias voltage dependent corrugation of the (8×2) superstructure. Close to $U = 0.8$ V the apparent corrugation inverts. Reprinted from [27], Copyright (1997), with permission from Elsevier

for the (8×2) structure. While the corrugation amounts to 0.45 ± 0.02 Å at $U = -0.8$ V it nearly vanishes at $U = 0.8$ V (being 0.03 ± 0.01 Å) and inverts for higher positive bias voltages.

A first indication of chain formation was found by Weller and Alvarado [22]. They report on eighth-order streaks in their LEED data along the [001] direction at a coverage $\Theta \approx 0.3$ ML. Kołaczkiewicz and Bauer [23] concentrated on the growth at submonolayer coverage at 300 and 1,200 K. At very low coverages ($\Theta \approx 0.15$ ML) they reported on stripes in the LEED pattern along the [001] direction. With increasing coverage the streaks soon resolve into spots of $(n \times 2)$ pattern with $n = 10, 7, 6, 5$ which are best resolved at $\Theta = 0.2, 0.28, 0.33$, and 0.4 ML, respectively [23]. Kołaczkiewicz and Bauer [23] explained these LEED pattern by Gd-row formation along the $[1\bar{1}0]$ direction which is driven by charge transfer from the electropositive adsorbate to the substrate. This results in a repulsive Coulomb interaction between adjacent gadolinium atoms. Since this interaction exhibits no intrinsic anisotropy the anisotropy of the W(110) substrate is of substantial importance for the formation of Gd chains. To explain the observations the repulsive Coulomb interaction must dominate along the [001] direction while it is overcompensated by an attractive interaction along the $[1\bar{1}0]$ direction.

Gd atoms adsorbed on Mo(110) also form chains but these are not linear as on W(110) but exhibit a zigzag form analogously with decreasing spacing between the chains as the coverage increases. The formation of these zigzag chain structures is accomplished due to indirect lateral interaction between adsorbed Gd atoms which can be presented as a screened Coulomb potential with account for Friedel oscillations [28]. The same behavior of forming zigzag chains was found for Nd atoms on Mo(110) [29].

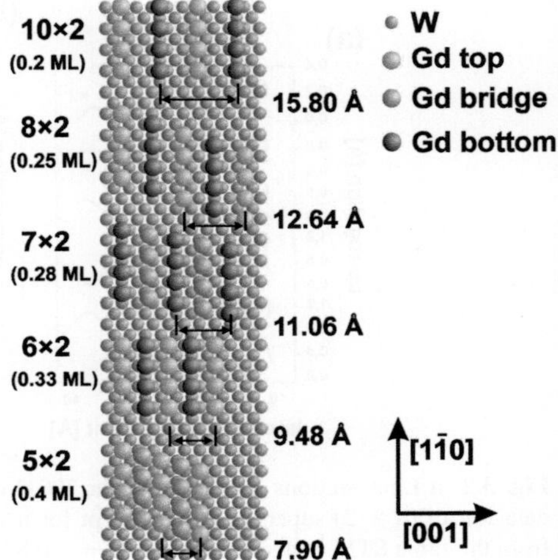

Fig. 3.3 Atomic structure models of the submonolayer superstructures formed by Gd/W(110)

Besides the (8 × 2) structure which has not been described before the LEED and STM results are in excellent agreement with an earlier work by Kołaczkiewicz and Bauer [23] who proposed a structure model which was extended to the (8 × 2) structure (see Fig. 3.3). Along the [001] direction the gadolinium adsorbates occupy alternating on-top and non-top sites. The latter are fourfold hollow sites for the (10 × 2), (8 × 2), and (6 × 2) structure and bridge sites in the case of the (7 × 2) and the (5 × 2) structure. Thus, chains and troughs in the STM images represent Gd atoms adsorbed at on-top and hollow sites, respectively. The measured periodicities along the [001] direction and the structure–model-based periodicities [in brackets] are: (10 × 2): (15.4 ± 0.5) Å [15.80 Å], (8 × 2): (12.6 ± 0.1) Å [12.64 Å], (7 × 2): (10.7 ± 0.5) Å [11.06 Å], (6 × 2): (9.2 ± 0.3) Å [9.48 Å], and (5 × 2): (7.8 ± 0.3) Å [7.90 Å].

As already shown in Fig. 3.2a the measured corrugation decreases with increasing Gd density. However, corrugations up to 1.2 Å as measured with the STM cannot solely be explained by the different adsorption topography since the corrugation of dense-packed metal surfaces is typically below 0.1 Å. In this context one should keep in mind that constant current STM images represent contour maps of the constant local density of states (LDOS). Therefore, scanning tunneling spectroscopy measurements have been performed by simultaneously recording topographic data and the normalized differential tunneling conductance $(dI/dU)/(I/U)$ as a function of the applied sample bias voltage U.

In Fig. 3.4 the tunneling spectra (a–e) are given corresponding to the superstructures (a–e) of Fig. 3.1. Since the normalization procedure leads to problems around zero bias the region from -0.2 to 0.2 V has been omitted. For the (8 × 2) and the (7 × 2) superstructures a distinction of tunneling sites above the Gd chains representing on-top (bridge) positions, respectively, and chains representing hollow positions are found. This was possible due to the relatively large inter-chain distance for these structures. The more dense structures did not reveal any

Fig. 3.4 Normalized tunneling spectra as measured for the five different superstructures already observed in the STM image of Fig. 3.1. *Black curves* correspond to tunneling sites above the maxima of the corrugation and the gray curves to tunneling sites above the minima, revealed at −0.8 V. Reprinted with permission from [26]. Copyright (1997) by the American Physical Society

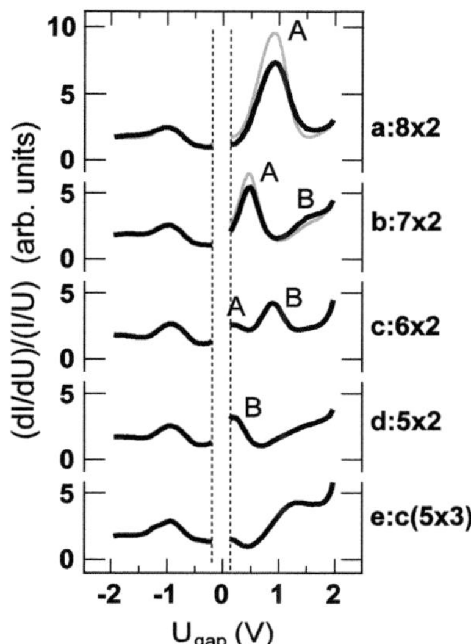

significant difference between these two sites. By comparing the tunneling spectra (a–e) obtained from the submonolayer coverage regime strong differences were observed, particularly peak positions as well as peak intensities of the empty states show a strong dependence on surface structure. A systematic trend in peak positions and intensities seems to be present by going from the (8×2) superstructure with a relatively large inter-chain distance to the (5×2) structure with the smallest inter-chain distance. In spectrum (a) a strong peak A at $U \approx 1$ eV is present. The strong peak in spectrum (b), labeled A, may be interpreted as the peak A of spectrum (a) shifted towards E_F due to the decreased inter-chain distance. This interpretation is supported by the observation that this trend continues for peak A as well as for peak B coming to the next closest structure, i.e. (6×2) in spectrum (c). Going from the (6×2) to the (5×2) peak B also shifts towards E_F. The spectrum (e) of the pseudohexagonal $c(5 \times 3)$ structure does not follow this trend, indicating that it exhibits different local electronic properties.

Additional information on the local electronic structure of the Gd films on W(110) in the submonolayer regime was obtained from spatially resolved measurements of the local tunneling barrier height which reflect spatial inhomogeneities in the local surface work function. Figure 3.5a shows a barrier height map of 0.5 ML Gd on W(110) which exhibits the same superstructures (a–e) already known from the STM image of Fig. 3.1. Additionally, a small surface area (f) with the structure of a close-packed first ML is observed. A detailed study concerning this close-packed layer is given in [30]. Six line sections of the different superstructures in the *dI/dz* map are displayed in Fig. 3.5b. Obviously, the tunneling barrier height and, therefore, the local work function decreases with decreasing inter-chain distance (a–d) by going from the (8×2) to the (5×2) superstructure. However, the next dense structure, i.e. the c (5×3), shows a distinctly higher

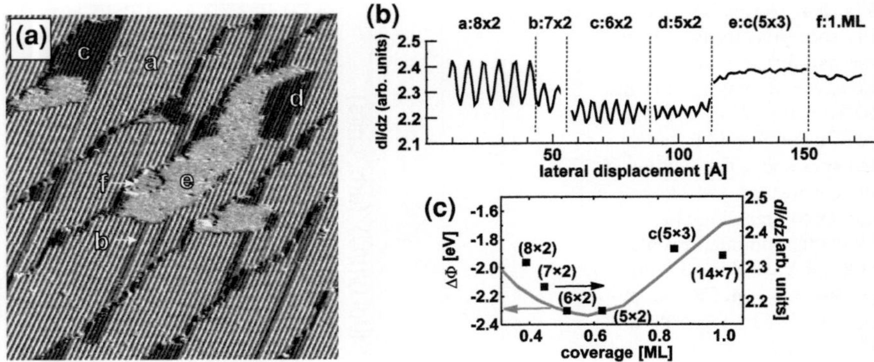

Fig. 3.5 a Spatially resolved map (80 nm × 60 nm) of the tunneling barrier height dI/dz reflecting spatial variations in the local surface work function (tunneling parameters: $U = -0.7$ V, $I = 1$ nA). The nomenclature used to mark the superstructures is the same as in Fig. 3.1. **b** Line sections of the dI/dz signal as measured above different superstructures. **c** The lowest barrier height is measured above the (5 × 2) and (6 × 2) superstructures, in agreement with work function measurements [23]. Reprinted from [27], Copyright (1997), with permission from Elsevier

work function compared to the (5 × 2) structure. As presented in Fig. 3.5c this STM based result on the nanometer scale is in excellent agreement with work function measurements reported earlier by Kołaczkiewicz and Bauer [23]. They observed the work function minimum to be associated with the maximum development of the (5 × 2) and (6 × 2) LEED pattern of a W(110) sample which Gd was continuously evaporated onto. While spatial averaging of different superstructures is inherent by using this method, the STM offers the possibility to directly observe the change of the local work function for each individual superstructure.

3.1.2 Morphologies of Epitaxial Gd(0001) Films

Previously to any investigation of electronic or magnetic properties of Gd(0001) on W(110) the influence of different preparation procedures on the Gd(0001) film morphology was systematically studied. Figure 3.6 shows constant current images of the three typical sample morphologies used in this study. The preparation procedure is based on an earlier publication [31]. The authors phenomenologically described the so-called "critical curve", indicating the thickness dependent annealing temperature which the transition from flat film growth to island formation occurs above. This dependence is schematically shown in Fig. 3.7.

Consistently with this work evaporation of (45 ± 5) ML Gd on the W(110) substrate held at 293 K and subsequent annealing at 710 K for 2 min leads to smooth Gd(0001) films (see Fig. 3.6a). The atomically flat terraces are 10 nm– 1 µm wide. Although these films reveal a sharp (1 × 1) LEED pattern and the

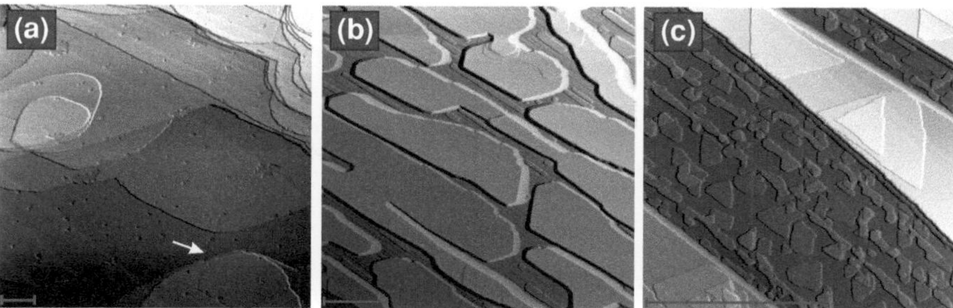

Fig. 3.6 Constant current topographs of typical sample morphologies used in this study: **a** Smooth Gd(0001) film grown on W(110) by evaporation of ≈ 45 ML at a substrate temperature of 293 K and subsequent annealing at 710 K for 2 min. The *arrow* points to a line defect on the surface. **b** Gd(0001) islands prepared by deposition of 10 ML on the substrate held at 530 K. Between the islands the substrate is covered by a distorted hexagonal monolayer. **c** Nucleation of second and third monolayer patches of Gd(0001) can be observed if 0.5 ML are deposited on a sample similar to (**b**) after cooling down to 293 K. Homoepitaxial growth occurs on the thick Gd islands. Every *scale bar* corresponds to 200 nm

Fig. 3.7 Schematic representation of the rare earth film morphology as a function of nominal film thickness and annealing temperature

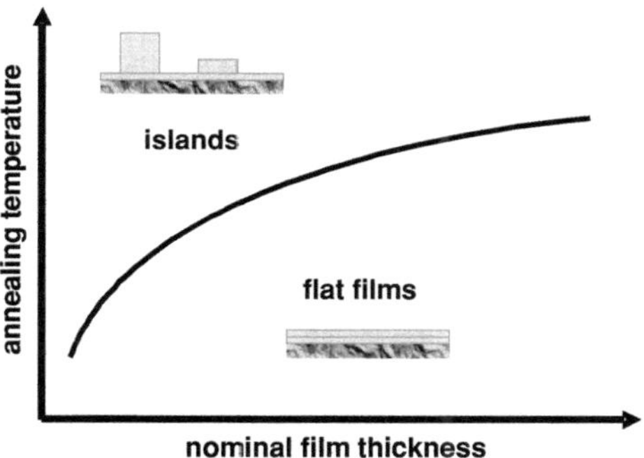

surface state is well-pronounced in photoemission spectra some line defects are visible on the surface (see arrow in Fig. 3.6a).

In contrast, island formation could be observed at lower coverage of approximately 10 ML even if the substrate temperature did not exceed 530 K (see Fig. 3.6b). These islands exhibit a local coverage of at least 4 ML, are atomically flat on top and will be described in more detail below. Between the islands the W(110) substrate is covered by a wetting layer of hexagonal but heavily distorted Gd. Two different models for the atomic structure of this so-called (14×7) structure have been described by other authors [23, 24].

Subsequent deposition of half a monolayer at room temperature on a sample similar to Fig. 3.6b results in samples with a large variety of local coverages. Due to the reduced mobility, nucleation of second and third layer patches occur on the strained monolayer of Gd/W(110), and homoepitaxial growth of Gd can be observed on Gd(0001)/W(110) (see Fig. 3.6c). Such a sample is favorable for a spatially resolving technique like STM/STS since it enables one to measure the surface electronic properties of all apparent coverages simultaneously in a single scan.

3.1.3 The Gd(0001) Surface State

It was one basic question in the framework of this investigation whether it is possible to observe the Gd(0001) surface state by means of STS. Figure 3.8a shows photoemission and inverse photoemission data of Gd(0001) published by Weschke et al. [4]. While the PE measurement shows the occupied part of the surface state at a binding energy $E_B \approx 100$ meV the empty part appears in the IPE spectrum at $E_B \approx -250$ meV. It is an advantage of STS to detect the contour of the local density of states on both sides of the Fermi level by tuning the applied voltage from negative to positive bias or vice versa.

Indeed, the dI/dU spectrum measured with the tip positioned above a Gd(0001) island (topography similar to Fig. 3.6b) exhibits two distinct maxima at a sample bias $U = -0.1$ and $U = 0.3$ V, being in good agreement to the binding energies

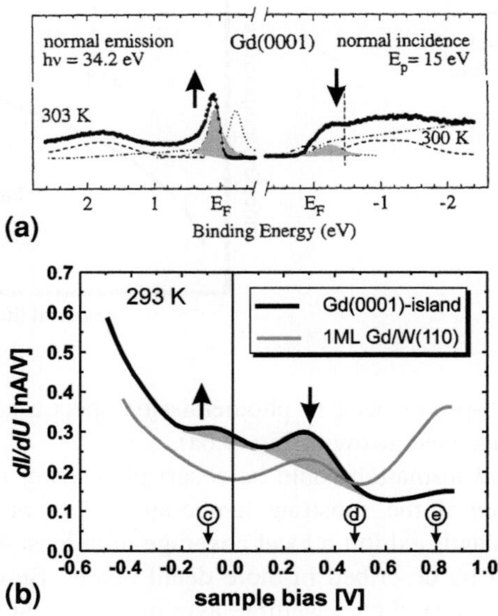

Fig. 3.8 **a** Photoemission (*left*) and inverse photoemission (*right*) spectra of Gd(0001) (from Weschke et al. [4]). The occupied part of the surface state appears in the PE spectrum (binding energy ≈ 100 meV) while the empty part is weakly visible in the IPE spectrum (binding energy ≈ -250 meV). **b** Tunneling spectrum measured on a sample similar to Fig. 3.6b above a Gd(0001) island and above the first distorted ML at 293 K showing both spin parts of the surface state. Spatially resolved data at sample bias values indicated by small *arrows* **c–e** will be shown in Fig. 3.10. From [33], copyright 1998, reproduced with permission from World Scientific Publishing Co. Pte. Ltd

for the occupied and the empty part of the surface state as determined by PE and IPE (cf. Fig. 3.8a), respectively. As described above a closed hexagonal but heavily distorted monolayer of Gd covers the W(110) substrate between the Gd islands. In contrast to the Gd(0001) islands the spectra measured above the monolayer exhibit only one asymmetric peak at a positive sample bias $U = 0.3$ V, i.e. in the empty sample states. Although the binding energy of this feature is identical to the empty part of the Gd(0001) surface state at room temperature it will be discussed in Sect. 4.1.2 that this peak does *not* represent a surface state as claimed in an earlier IPE study [32].

It is a typical property of a surface state that it vanishes upon exposure of the clean surface to even small amounts of adsorbates. It was checked that both peaks in the STS spectra that are characteristic for the surface state can be quenched by hydrogen, oxygen, or carbon monoxide down to exposures of approximately 1 L (1 Langmuir $= 1 \times 10^{-6}$ torr $\cdot$ s) [34]. Figure 3.9a shows the topography of Gd(0001) islands after exposure to 0.2 L hydrogen. While the surface of island A is homogeneous two different heights can be recognized on island B. Some protrusions (e.g. arrow C) appear about 1.5 Å higher than the surrounding island surface (arrow D). Typical tunneling spectra measured above both parts of this inhomogeneous Gd island are shown in Fig. 3.9b. The double peak structure being characteristic for the clean Gd(0001) surface state can be recognized in the spectra measured above the protruding areas (arrow C). In contrast, the surrounding island surface exhibits a much lower dI/dU signal at low positive and negative sample

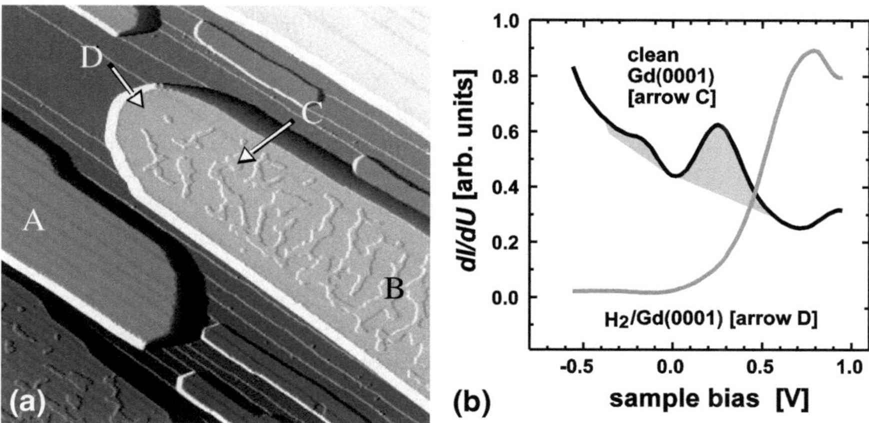

Fig. 3.9 **a** Topography of Gd(0001) islands after exposure to hydrogen measured at a sample bias $U = -0.3$ V. The surface of island A is homogeneous. Some protruding areas (*arrow C*) can be recognized on island B with an apparent height being about 1.5 Å higher than the surrounding island surface (*arrow D*). **b** Tunneling spectra reveal that peaks being characteristic for the surface state are quenched at locations with a reduced apparent height (**d**). In order to keep the tunneling current constant the tip has to approach closer towards the sample wherever hydrogen has been adsorbed than on clean Gd(0001), consequently clean Gd appears protruding. From [33], copyright 1998, reproduced with permission from World Scientific Publishing Co. Pte. Ltd

bias indicating a strongly reduced density of states around the Fermi level caused by the local adsorption of hydrogen. A closer inspection of the Gd islands reveals that most of them remain clean, e.g. island A, while some are contaminated by hydrogen as island B. This is a consequence of the fact that adsorption preferably starts at surface imperfections while defect-free islands remain unaffected at low hydrogen partial pressure. The behavior of adsorbates will be discussed in more detail in Chap. 4.

3.1.4 Thickness and Morphology Dependent Electronic Properties

The surface state appears in the dI/dU spectra as a double peak structure which can be quenched by the local adsorption of hydrogen. In the following, it will be discussed whether or not a dependence of the differential conductance dI/dU on the island thickness exists. Figure 3.10a shows the topography of a sample prepared by deposition of 5 ML Gd/W(110) at 530 K resulting in Stranski–Krastanov growth, i.e. island formation. Two islands can be recognized. The surface of both islands appears atomically flat. Below the islands the substrate exhibits several monoatomic steps. Therefore the coverage increases by going from the right to the left island edge as schematically indicated in the line section (see Fig. 3.10b). The local coverages amount to 7 ML $\leq \Theta_{loc} \leq$ 19 ML and 4 ML $\leq \Theta_{loc} \leq$ 22 ML for the island in the upper and lower part of the image, respectively. Since the interlayer spacing $d_{Gd}(0001) = 2.89$ Å exceeds $d_W(110) = 2.23$ Å the substrate miscut is overcompensated and the Gd islands exhibit a slope although they are atomically flat on top.

Simultaneously with the topography a dI/dU spectrum was determined at every pixel of the scan. During the measurement the sample was held at $T = 293$ K. Figures 3.10c–e show maps of the differential conductance dI/dU for different sample bias: (c) $U = +0.8$ V, (d) $U = +0.47$ V, and (e) $U = -0.1$ V. These sample bias values have been marked by arrows at the bottom axis of Fig. 3.8b. The differential conductance is gray coded, i.e. the higher the local dI/dU signal the brighter a location appears. At a sample bias $U = +0.8$ V the tunneling current is dominated by electrons which tunnel from the tip into unoccupied sample states with a binding energy of +0.8 eV. Comparison with the topographic data of Fig. 3.10a reveals that at this particular binding energy the differential conductance above the Gd monolayer is higher than above any island. Beside a few small bright spots the dI/dU signal at $U = +0.8$ V measured above the Gd island is uniform and therefore independent of the local coverage in the range 4 ML $\leq \Theta_{loc} \leq$ 22 ML.[1] In the dI/dU maps no contrast was found on the

---

[1] The bright spots reflect an increased LDOS at this particular sample bias which is induced by the local adsorption of hydrogen and which has already been described in Fig. 3.9b. However, it should be mentioned that the total amount of hydrogen adsorbed on the surface is far less than 0.01 L.

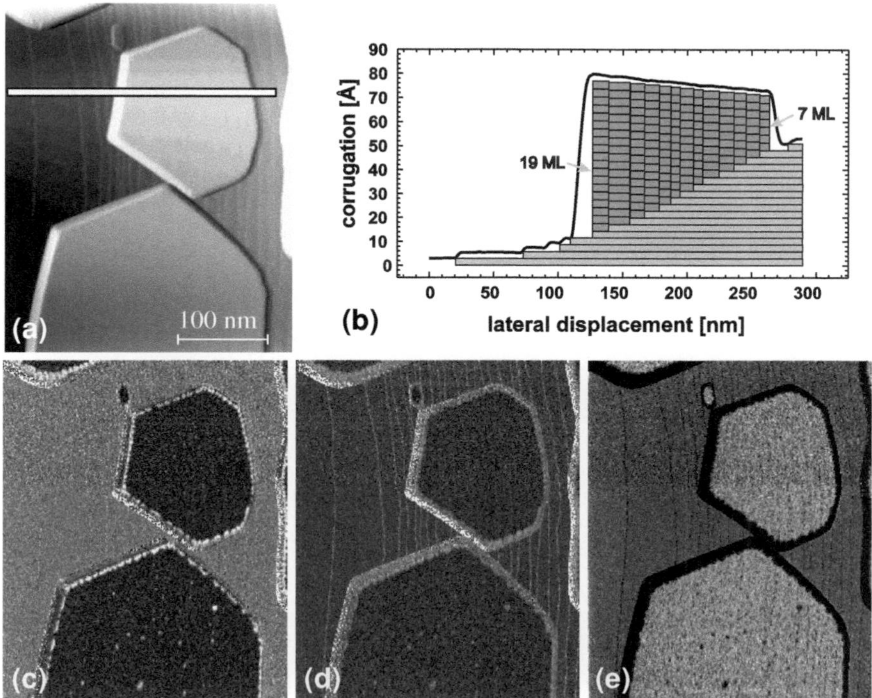

Fig. 3.10 a Constant-current STM topograph of nominally 5 ML Gd(0001)/W(110). The substrate was held at 530 K during evaporation resulting in Stranski–Krastanov growth. **b** Line section extracted from **a**. Since the substrate exhibits steps below the atomically flat islands the local coverage of the Gd island increases from 7 ML up to 19 ML by going from the right to the left. Maps of the differential conductance dI/dU measured at different sample bias are shown in **c** $U = +0.8$ V, **d** $U = +0.47$ V, and **e** $U = -0.1$ V. The dI/dU signal and therefore the local density of states does not depend on the local coverage. From [33], copyright 1998, reproduced with permission from World Scientific Publishing Co. Pte. Ltd

Gd island at any sample bias (cf. Fig. 3.10c–e) in the voltage range under study $(-0.6$ V $\leq U \leq +0.9$ V). At $U = +0.47$ V the contrast between the island surface and the monolayer vanishes. Comparison with the spectra of Fig. 3.10b reveals that at this sample bias the dI/dU signal of the Gd monolayer is equal to Gd(0001) islands. The contrast inverts if the sample bias is further reduced. For instance Fig. 3.10e shows a map of the dI/dU signal at $U = -0.1$ V, i.e. close to the energetical position of the occupied surface state.

It was shown until now that the electronic structure which is manifested in the dI/dU spectra of clean Gd(0001) islands, i.e. the energetical position of the surface state and its intensity, does not dependent on the local coverage for $\Theta_{\mathrm{loc}} \geq 4$ ML. It is known, however, that the first monolayer of Gd/W(110) does not exhibit the surface state (cf. Fig. 3.8b). Consequently, the question arises: What is the critical thickness for the surface state to appear in the tunneling spectra? To unravel this

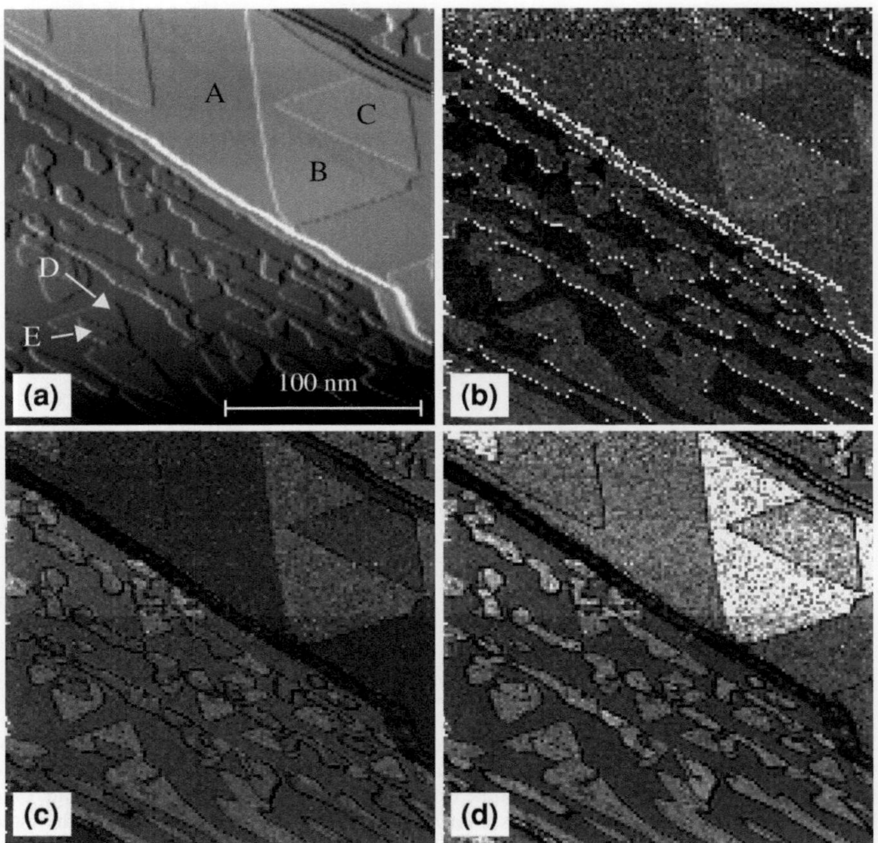

Fig. 3.11 a Constant-current STM image of a sample surface similar to that shown in Fig. 3.6b simultaneously exhibiting a large variety of local coverages. On top of the high Gd(0001) island (*A*) triangular shaped Gd islands (*B*, *C*) appear. Nucleation of second (*D*) and third (*E*) monolayer patches can be observed on the first monolayer of Gd/W(110). Maps of the differential conductance dI/dU measured at different sample bias are shown in **b** $U = +0.4$ V, **c** $U = -0.15$ V, and **d** $U = -0.2$ V. The sample temperature during the measurement was 117 K. From [33], copyright 1998, reproduced with permission from World Scientific Publishing Co. Pte. Ltd

problem tunneling spectroscopy was performed on a sample with a morphology similar to Fig. 3.6c prepared at room temperature deposition of 0.5 ML Gd on a Stranski–Krastanov film of Gd/W(110). Figure 3.11a shows a Gd(0001) island with label A. On top of the island surface homoepitaxial growth of triangular shaped Gd islands (e.g. B and C) can be observed. Second (D) and third (E) monolayer patches have nucleated on the first monolayer. The smallest patches exhibit an island area of approximately 25 nm^2.

Again, the differential conductance was measured simultaneously with the topography. Figure 3.11b–d shows maps of the dI/dU signal at three particular bias voltages: (b) $U = +0.4$ V, (c) $U = -0.15$ V, and (d) $U = -0.2$ V. It is a striking fact that island B exhibits a *dI/dU* signal which differs from the underlying island A best visible in Figs. 3.11c and d. In total it was found that approximately 10% of all triangular islands grown at 300 K exhibit this property. In contrast, island C which is grown on top of island B does not exhibit a significantly different *dI/dU* signal compared to island A. One possible explanation for this change of the electronic structure with respect to the underlying (0001) surface is a stacking fault between the surface and the first subsurface layer. It is also obvious from the spatially resolved data that the electronic structure of second and third monolayer patches differ from each other and that both differ from the Gd monolayer (note to the sites marked by arrows D and E in Fig. 3.11a).

The difference becomes obvious in the tunneling spectra measured at sites with different local coverages as plotted in Fig. 3.12. To prevent an unwanted overlap the spectra have been shifted relative to each other. As already described for room temperature measurements (cf. Fig. 3.8b) the Gd monolayer exhibits an asymmetric peak centered at a sample bias $U = +0.3$V which does not represent a surface state. In contrast, the spectrum measured above a double layer patch shows two maxima of the *dI/dU* signal: a distinct peak at $U \approx +0.45$ V and a weak shoulder at $U \approx -0.1$ V. Similar to the results previously shown in Fig. 3.9 experiments were performed which show that both features vanish upon contamination (not shown here).

Therefore it can be concluded that the surface state already exists on Gd patches with a thickness of two atomic layers and down to an area of at least 25 nm^2. If the local coverage is increased the occupied part of the surface state shifts to a higher binding energy. For $\Theta_{loc} = 3$ ML it amounts to -140 meV until it converges for $\Theta_{loc} \geq 4$ ML to the thick film limit of -180 meV. The binding energy of the empty part of the surface state exhibits a much weaker dependence

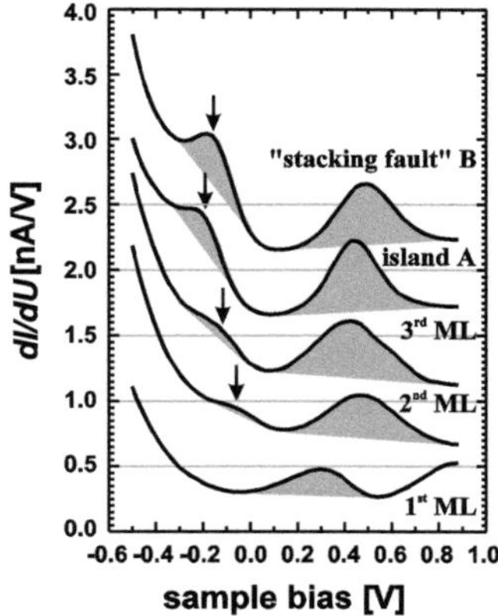

Fig. 3.12 Tunneling spectra measured at $T = 117$ K for local coverages $\Theta_{loc} = 1, 2, 3$, and ≥ 4 ML ("island"). For local coverages of 2 and 3 ML the surface state exists but the exchange splitting is reduced. The spectrum of island B in Fig. 3.11a is shifted by about 25 meV, possibly due to a stacking fault which will be discussed below. From [33], copyright 1998, reproduced with permission from World Scientific Publishing Co. Pte. Ltd

Fig. 3.13 Model of a stacking fault on an hcp(0001) surface. The stacking sequence *ABAB*... is replaced by *ABC* if a stacking fault occurs

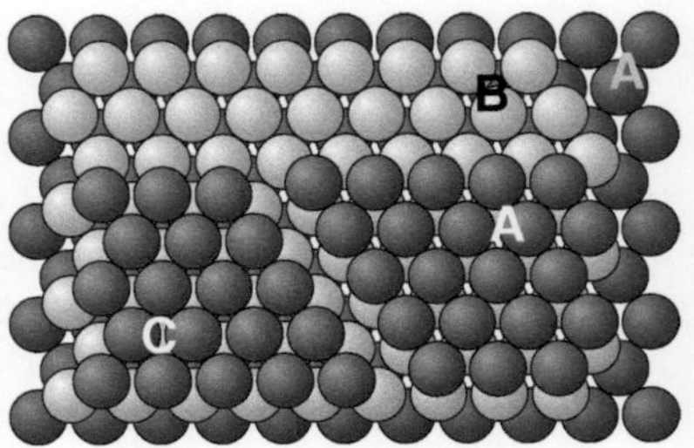

on the local coverage. This behavior points to surface electronic properties being fully developed for films with a thickness of about four monolayers. The most likely explanation for this behavior is that the surface stress of epitaxially grown Gd(0001)/W(110) decreases with increasing coverage thereby approaching the equilibrium lattice constant of an unstrained hcp(0001) surface. This explanation qualitatively agrees with an earlier LEED study performed by Weller and Alvarado [22]. According to this reference the initial Gd surface stress ($8 \pm 3\%$ at $\bar{d} = 3$ Å) reduces by $\approx 75\%$ by increasing the coverage to ≈ 4 ML and is fully relaxed at ≈ 30 ML.

The question arises whether it is possible to detect *unambiguously* stacking faults using STS. Figure 3.13 presents a model for the stacking sequence of an hcp(0001) surface (*ABAB*...; right) and for a stacking fault (*ABC*; left). For this purpose a sample with a lot of terraces was prepared. This type of sample can be obtained by a preparation which starts at elevated temperature (≈ 400 K). After 1 min heating has been stopped allowing the substrate temperature to slowly decrease during the subsequent further evaporation. The decreasing mobility of rare earth metal atoms results in pyramidally arranged terraces with a (0001) orientation at the surface. A comprehensive description concerning the growth of ultrathin metal films is given in [35].

Figure 3.14a shows the surface of another rare earth metal Tb being prepared by this method. Comparative measurements demonstrated that no difference to Gd occurs. More than 90% of the surface exhibits a differential conductance which is shown in Fig. 3.14b as the black curve and is presumed to be the defect-free Tb(0001) surface. However, $\approx 7\%$ of the area shows *dI/dU* spectra given as the gray curve with an enhanced differential conductance at negative and low positive bias voltages. These terraces are predicted to possess stacking faults. Figure 3.14c shows a map of the differential conductance at about 0 V. The bright appearing terraces are due to the increased differential conductance. This area of the sample is marked by the larger white box in Fig. 3.14a. The observation due to this map that in each case the whole terrace exhibit a uniform spectroscopy curve is a further hint to stacking faults as the reason for the difference in the electronic properties.

A careful inspection results in the observation that a change in the differential conductance also occurs within one terrace. This is demonstrated in Fig. 3.15. This location on the sample corresponds to the smaller white box in Fig. 3.14a. The map of the differential conductance exhibits a contrast on top of a terrace the

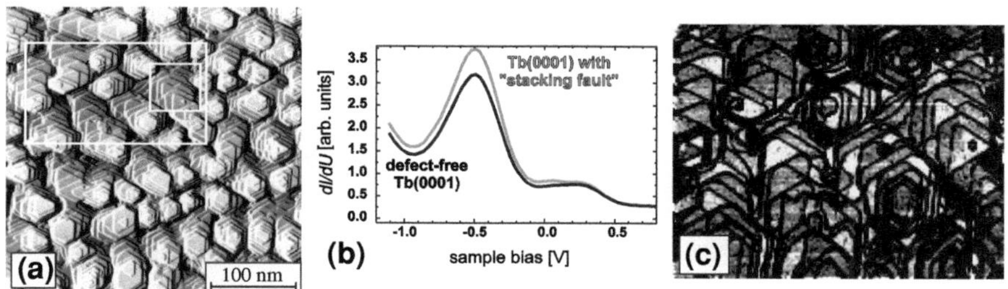

Fig. 3.14 **a** Morphology of a Tb sample being prepared in such a way that a lot of terraces occur. The detailed preparation procedure is described in the text. **b** Most of the terraces exhibit a differential conductance which is shown as the *black curve*. But a few of them show *dI/dU* spectra given as the *gray curve*. **c** Map of the differential conductance at about 0 V. The bright appearing terraces are due to the increased differential conductance. This area of the sample is marked by the larger white box in **a**

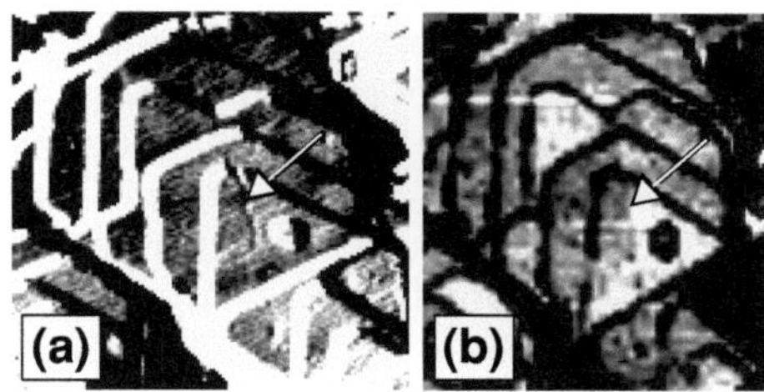

Fig. 3.15 **a** Topography (60 nm × 60 nm) of that location which is marked by the small *white box* in Fig. 3.14. This image is differentiated in order to enhance the contrast. A dislocation line on a terrace is marked by the *arrow*. **b** Corresponding map of the differential conductance at a bias voltage of 0 V. The change in brightness, due to the variation of the differential conductance, at exactly the position marked by the arrow gives evidence for a stacking fault where the dislocation line occurs

boundary of which is marked by the arrow (see Fig. 3.15b). If this is really caused by a stacking fault then it is necessary that a dislocation occurs following exactly this line. The topography of the corresponding area (see Fig. 3.15a) demonstrates indeed this behavior (see arrow). The occurrence of the dislocation line becomes understandable by Fig. 3.13 since a dislocation line must occur between both areas labeled A and C. Using the STM in the spectroscopic mode therefore enables to detect the occurrence of stacking faults on crystalline surfaces.

3.2 Binary Alloys of Rare Earth and Transition Metals

In this section the determination of the growth behavior of thin $GdFe_2$ films on W(110) is presented. It will be shown that an epitaxial growth of up to two monolayer thick films is possible. Based on the atomically resolved STM images and LEED data a structure model for ultrathin $GdFe_2$ films is proposed and compared to the $GdFe_2$ bulk structure (Laves phase C15).

3.2.1 Remarks on the Preparation Procedure

The rare earth metal Gd was evaporated from a W crucible heated by electron bombardment whereas Fe was evaporated from a rod. The evaporation flux was repeatedly calibrated with submonolayer accuracy via STM by growing pure films of Gd or Fe. Consequently, the amount of evaporated metal could be ascertained by about ± 0.2 ML for the very thin films of $GdFe_2$. The exact amount was determined afterwards via STM. All topographic STM images were measured in the constant-current mode. The scanner was calibrated on the well known Gd/W(110) superstructures [23, 26, 27]. The radii of Gd and Fe significantly differ from each other. The first ML of Fe on W(110) that grows pseudomorphically holds 1.56 times the number of adsorbate atoms compared with the first ML Gd/W(110). For this reason, the amount of evaporated material is given in substrate units to preserve the mixing ratio to be directly comparable. The first closed monolayer of Gd on W(110) holds 0.64 monolayers with respect to substrate units; the first closed monolayer of Fe on W(110) that grows pseudomorphically consequently holds 1 monolayer in substrate units.

3.2.2 Submonolayer Coverage Regime

Both elements, Fe and Gd, exhibit Frank–van der Merwe or Stranski–Krastanov growth depending on the film thickness and annealing temperature in the coverage range above one monolayer [22, 36, 37]. The thin film growth behavior of the two metals in the coverage regime below one monolayer differs drastically from each other. This is demonstrated in Fig. 3.16 where two samples of both metals in pure form, prepared under similar conditions, are presented. While Fe (see Fig. 3.16a) exhibits the formation of one monolayer film patches as well as stripes of material growing along the substrate steps (step flow growth) [36, 37], the Gd atoms (see Fig. 3.16b) tend to cover the W(110) surface as quasi-one-dimensional superstructures [23, 26] (see Sect. 3.1.1).

Given these two elements, Gd and Fe, differing drastically from each other in their thin film growth behavior on W(110), the question arises which growth mode an alloy of the two components will show and which stoichiometry is the most stable one on top of the W(110) surface. Figure 3.17 shows a topographic STM image of an alloy of 0.3 ML Gd and 0.4 ML Fe. The sample was prepared by evaporating 0.3 ML Gd onto the substrate held at 400 K and subsequently by an additional evaporation of 0.4 ML Fe followed by a final annealing at 700 K for 5 min. The substrate is covered by a film consisting of two different kind of domains, a striped one with the stripes running along the $[1\bar{1}0]$ direction of the

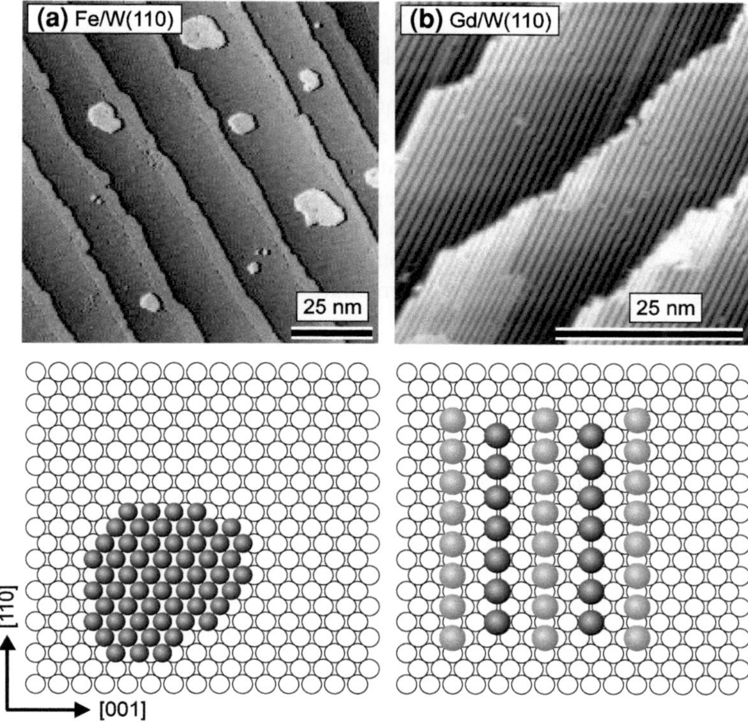

Fig. 3.16 Comparison of the growth of **a** Fe and **b** Gd on W(110) in the submonolayer coverage regime. For both metals the coverage is about 0.25 ML. The scan range for both images is 70 nm × 70 nm. Below the STM images structure models are shown to highlight the difference in the growth mode. Reprinted with permission from [38]. Copyright (1999) by the American Physical Society

Fig. 3.17 Constant current STM image of an alloy of approximately 0.3 ML Gd and 0.4 ML Fe. The *striped areas* represent the well-known Gd superstructures with the stripes being aligned along the [001] direction of the substrate. The smooth areas correspond to an alloy of GdFe$_2$. The scan range is 70 nm × 70 nm. Sample bias: $U = 0.2$ V, tunneling current $I = 0.3$ nA. Reprinted with permission from [38]. Copyright (1999) by the American Physical Society

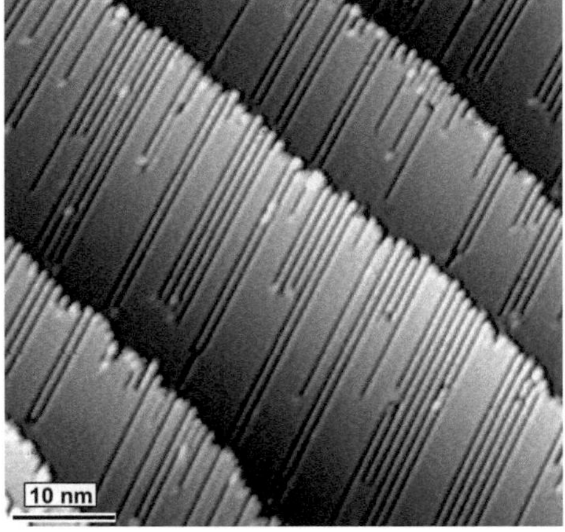

substrate being typical for Gd, as well as a smooth one. This sample exhibits a LEED pattern labeled as $\begin{pmatrix} 2 & 1 \\ 1 & 2 \end{pmatrix}$ not known from pure Gd or Fe films. In view of the mixing ratio and the assumption that the striped areas consist mainly of Gd, the smooth areas are expected to consist of an alloy with a 1:2 mixing ratio of Gd to Fe.

To strengthen this hypothesis, a sample was prepared that holds 0.3 ML of Gd and 0.6 ML of Fe. The preparation steps as well as the annealing procedure was the same as for the sample displayed in Fig. 3.17. Figure 3.18a shows the resulting sample topography. The substrate is completely covered by a smooth film with the underlying W(110) substrate steps remaining visible. The LEED pattern of this sample corresponds to a sharp $\begin{pmatrix} 2 & 1 \\ 1 & 2 \end{pmatrix}$ superstructure. A photograph of the LEED pattern as well as its interpretation is presented in the Fig. 3.18c and d, respectively. The atomic distances as determined from the diffraction pattern are 9.48 Å in the [001] direction and 4.47 Å in the $[1\bar{1}0]$ direction of the substrate.

The same periodicity as in the LEED pattern can be found in atomically resolved STM images of the first ML GdFe$_2$ alloy. Figure 3.18b shows a part of the sample presented in Fig. 3.18a with atomic resolution. Based on the LEED and STM data as well as on the stoichiometry of the prepared films a structure model for the alloy is presented in Fig. 3.19. All atoms are drawn to scale of their metallic radii. In this model the Gd as well as the Fe atoms have been placed on bridge sites with respect to the underlying W(110). This is assumed to be the energetically favorable adsorption site for both atomic species since it was shown that this is true for Fe on W(110) [24].

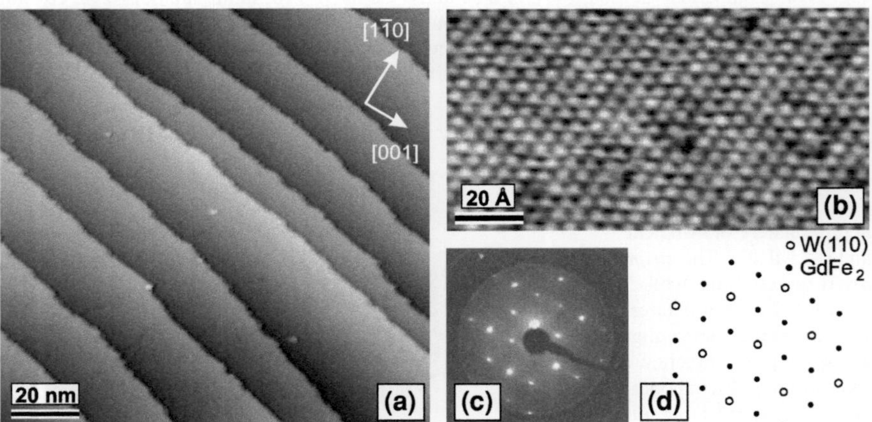

Fig. 3.18 a Completely closed and smooth first ML of GdFe$_2$ on W(110). **b** Atomic resolution obtained on this sample at a sample bias of $U = 0.18$ V and a tunneling current of $I = 3$ nA. **c** Photograph and **d** sketch of the $\begin{pmatrix} 2 & 1 \\ 1 & 2 \end{pmatrix}$ LEED pattern of this sample. The crystallographic directions are the same for all figures. Reprinted with permission from [38]. Copyright (1999) by the American Physical Society

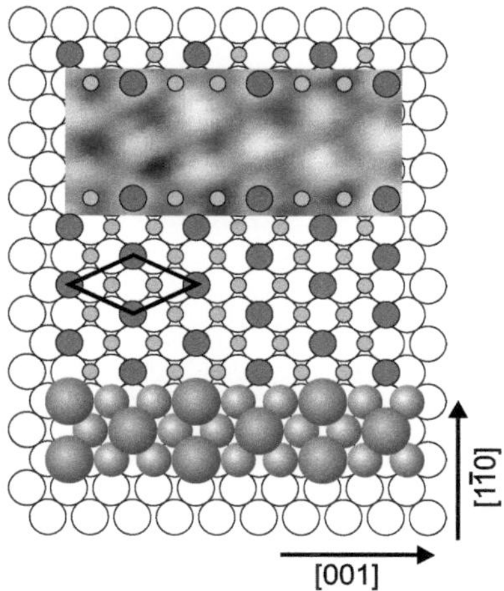

Fig. 3.19 Structure model for the first ML GdFe$_2$ on W(110). Gd is represented as large, Fe as small balls. The atoms are scaled down by a factor of two for better clarity of the registry between substrate and adsorbate. In the lower part of the model the atoms are drawn to scale. A part of an atomically resolved STM image is inserted in the structure model at the same scale. Reprinted with permission from [38]. Copyright (1999) by the American Physical Society

Comparing the structure model with the atomic scale STM images it becomes clear that one does only see one atomic species of the alloy, the Gd atoms. The fact that the Gd atoms are much larger than the Fe atoms may play a major role but a definite statement concerning this point can only be given if LDOS calculations for this structure will become available.

3.2.3 Structural Determination as Function of Film Thickness

The growth of the GdFe$_2$ structure could be continued to the second ML by evaporating approximately 0.5 ML of Gd and 1 ML Fe and subsequent annealing to 500 K [38–40]. The resulting sample topography is shown in Fig. 3.20. The substrate is

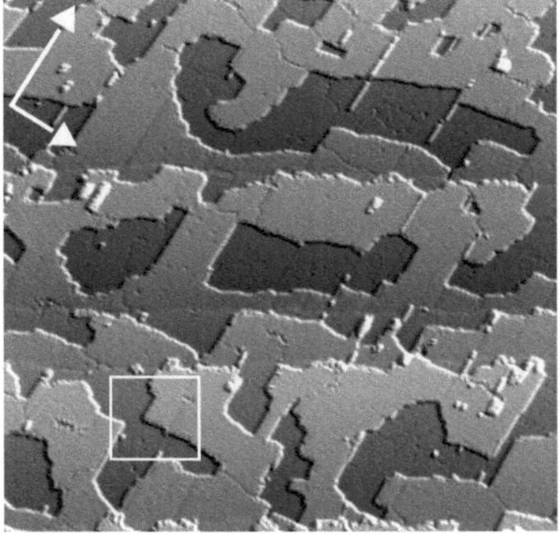

Fig. 3.20 First and second ML GdFe$_2$ on W(110). The scan range is 100 nm $\times$ 100 nm. In the area marked with a box atomic resolution is obtained (see Fig. 3.21). Reprinted from [39], Copyright (1999), with permission from Elsevier

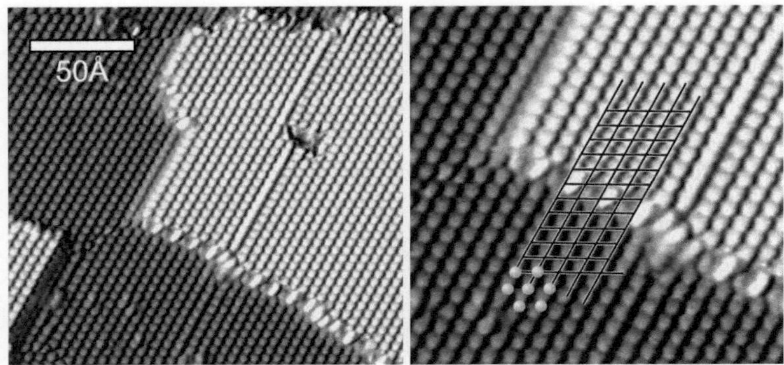

Fig. 3.21 Atomic resolution on first and second ML GdFe$_2$ on W(110) obtained at $U = 55$ mV and $I = 3$ nA. The *right* image is enlarged from the *left* one in order to obtain a direct comparison of the locations in both layers. The structure model for the first and second ML GdFe$_2$ on W(110) basing on the atomically resolved data is shown in Fig. 3.22. Reprinted from [39], Copyright (1999), with permission from Elsevier

completely covered by a smooth alloy of Gd and Fe with patches of the second ML exhibiting the stoichiometry of GdFe$_2$. The samples that were prepared in this way gave a LEED pattern that showed a sharp $\begin{pmatrix} 2 & 1 \\ 1 & 2 \end{pmatrix}$ superstructure, not known from clean Fe or Gd on W(110). The arrows are aligned along the $[1\bar{1}0]$ direction of the substrate (long arrow) and the [001] direction (short arrow). Due to the steps of the underlying substrate the height increases from the top to the bottom but on each terrace the GdFe$_2$ film is only one or two monolayers thick.

In the area marked by a box, atomic resolution could be achieved on top of the first and the second ML simultaneously. This image is presented in Fig. 3.21.

The same periodicity as in the LEED pattern can be found in the atomically resolved STM images. Based on these data a structure model for the alloy is deduced and shown in Fig. 3.22. This structure was the only stoichiometry to fit the data. All atoms are drawn to scale of their atomic radius.

Figure 3.23 displays the tunneling spectroscopy measurements that were carried out on this sample. The differential conductance dI/dU is a direct measure of the local density of states. Within the error of the measurement there is no difference between the first and the second ML of GdFe$_2$. This observation reflects the identical geometric arrangement of the alloy in both layers.

With images showing atomic resolution on top of the first as well as on the second ML, it is possible to investigate the positions of the atoms in the second ML with respect to the atoms in the first ML. This may give an idea about the crystallographic structure of the sample. To find the positions of the second ML maxima visible in the STM images in comparison to the first ML maxima, the periodicity of the second ML is continued above the first ML without crossing dislocations. The grid shown in Fig. 3.21 is the result of such a procedure; the position of the second ML maxima with respect to the first ML is additionally

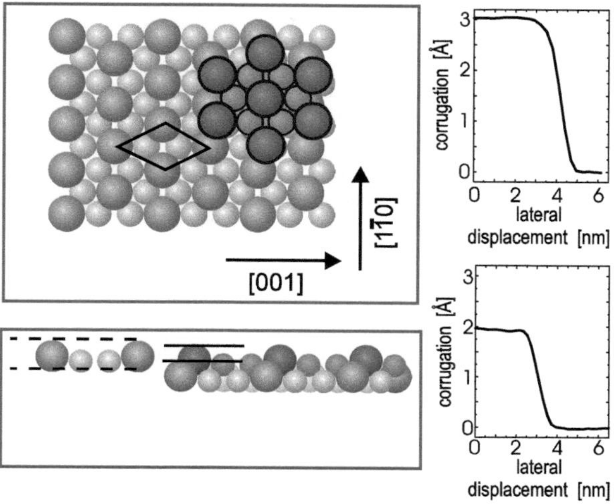

Fig. 3.22 *Top view* and *side view* of the two ML high GdFe$_2$ film. The line section for the first ML (*top*) from Fig. 3.24 and the second layer (*bottom*), taken from Fig. 3.21, corroborates the deduced structure. Reprinted from [39], Copyright (1999), with permission from Elsevier

marked with balls. Again it is assumed that the maxima visible in the atomically resolved STM images represent the Gd atoms.

The resulting model in top as well as in side view is presented in Fig. 3.22. The step is only shown in order to compare directly the different step heights. Assuming that this structure model is correct, one should observe a distinctly larger step height from the W(110) substrate to the first ML than from the first ML to the second one being about 2 Å (see line sections in Fig. 3.22).

Indeed, this was observed after succeeding in preparing the first ML GdFe$_2$ covering the substrate only partially. An image of this sample is shown in Fig. 3.24. The substrate was held at room temperature during evaporation of 0.25 ML Gd. After the evaporation of 0.5 ML Fe, the sample was subsequently annealed at 700 K for 5 min. The one ML thick GdFe$_2$ film grows along the substrate steps leaving the W(110) substrate partially uncovered. Some rectangular holes in the film being oriented along the main crystallographic directions of the substrate are present. The line sections B and C demonstrate that the uncovered

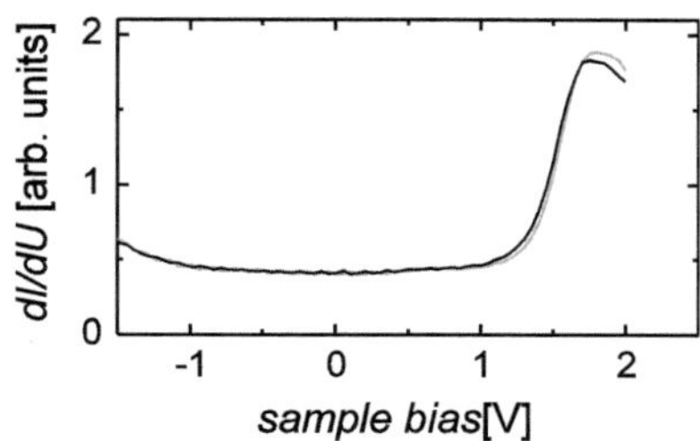

Fig. 3.23 Differential conductance *dI/dU* obtained on the first (*gray*) and second (*black*) ML GdFe$_2$ on W(110). The stabilization parameters are $U = 2$ V, $I = 0.5$ nA. Reprinted with permission from [38]. Copyright (1999) by the American Physical Society

Fig. 3.24 STM image of 0.8 ML GdFe$_2$ with a small Gd surplus growing in step flow along the W(110) steps edges. Rectangular holes in the alloy film are oriented along the [001] and [1$\bar{1}$0] directions of the substrate. The line sections B and C demonstrate that the uncovered tungsten substrate is visible on the bottom of the holes. The tunneling parameters are $U = -0.24$ V and $I = 0.2$ nA. The sections were obtained along the lines shown in the image. Reprinted from [39], Copyright (1999), with permission from Elsevier

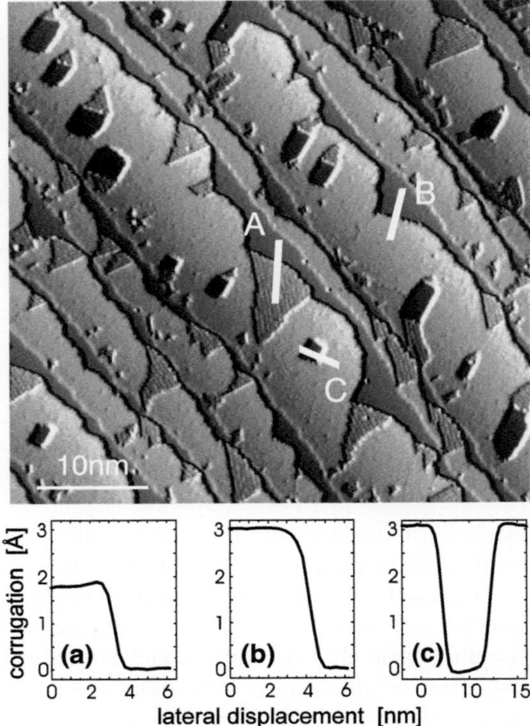

tungsten substrate is visible on the bottom of the holes. A slight Gd surplus is present in small, triangular shaped areas of the first ML of Gd that can be identified by the typical dislocation network [26]. The step height (see line section A) is smaller than for the alloy. The line sections show that the monoatomic step height of 3 Å for the first ML is much larger than 2 Å for the second one (cf. Fig. 3.22). This observation independently corroborates the structure model for the first and second ML.

In the next step the structure model for ultrathin GdFe$_2$ films is compared with the known crystallographic structure of bulk GdFe$_2$, the so-called C15 Laves phase being a complicated arrangement of the two atomic species (for further details see, e.g., Ref. [41]). No crystallographic directions with low indices of the Laves phase are found corresponding to the model which was deduced as a result of the STM and LEED data. This is not surprising because the first as well as the second ML grows pseudomorphically, i.e. the atomic arrangement in the alloy is mainly determined by the substrate, resulting in a relatively large deviation of the atomic positions when comparing the deposited film and the bulk Laves phase.

Nevertheless, the geometric arrangement of the Gd atoms in the GdFe$_2$ films of up to two ML thickness appears to resemble the (111) plane of the C15 Laves phase, compressed by 14% in the [1$\bar{1}$0] direction of the W(110) substrate and strained by 5.3% in the [001] direction. However, the arrangement of the Fe atoms with respect to the Gd atoms is completely different from the bulk structure. Assuming a release of this deviation with growing film thickness it can be expected that the C15 Laves phase develops at larger film thickness of $GdFe_2$ on W(110) as deduced for $TbFe_2$ on Mo(110) [15]. This would imply the existence of a structural phase transition in the $GdFe_2$ system.

Increasing the amount of Fe additionally enables to prepare $GdFe_3$ layers on W(110) [42].

3.3 Heavy Fermion Systems as Ternary Intermetallics with Extraordinary Properties

Heavy Fermion Systems (HFS) are particular metallic materials which contain so-called "heavy" electrons. The characteristic observation for these systems is that the electrons which are responsible for the electric transport possess a high effective mass. This phenomenon, however, arises at low temperatures of about 10 K. At high values, e.g. room temperature, HFS show a "normal" behavior.

Heavy Fermion Systems are intermetallics which consist of rare earths or actinides together with other metal species. Examples are $CeAl_3$ [43] and UPt_3 [44]. These materials have partially filled $4f$ or $5f$ shells. At high temperatures the f electrons are localized. This behavior is comparable to "conventional" alloys with rare earths or actinides. With decreasing temperature the systems order in an antiferromagnetic state. Heavy Fermion Systems, however, behave like normal metals but the effective mass of the electrons is significantly enhanced (often by a factor of hundred).

Electron spectroscopic techniques enable to determine the electronic properties in a very direct way. But, *atomically clean* and *well-ordered* surfaces are necessary for this investigation due to the surface sensitivity. Therefore, at the beginning it will be demonstrated that the evaporation of HFS on W(110) results in samples which are suited for electron spectroscopy [45–47].

3.3.1 Preparation of Well-ordered Surfaces

The left part of Fig. 3.25 shows a LEED pattern of $CePd_2Si_2$ on W(110) at an energy of 58 eV. The film with a thickness of about ten layers was prepared by evaporation at room temperature with a subsequent annealing at 1,000 K for 30 min. The spot rows are oriented along the $\langle 110 \rangle$ direction of the W(110) crystal. Every third spot on the lines has higher intensity as the other ones and every second line of spots is much weaker pronounced. The crystal structure of $CePd_2Si_2$ is shown in the right part of Fig. 3.25, it is of the type RT_2Y_2 [48]. The lattice constants are $a = 4.239$ Å and $c = 9.889$ Å. The sites in the given elementary cell are for Ce: (0, 0, 0), for Pd: (1/2, 0, 1/4), and for Si: (0, 0, z) with

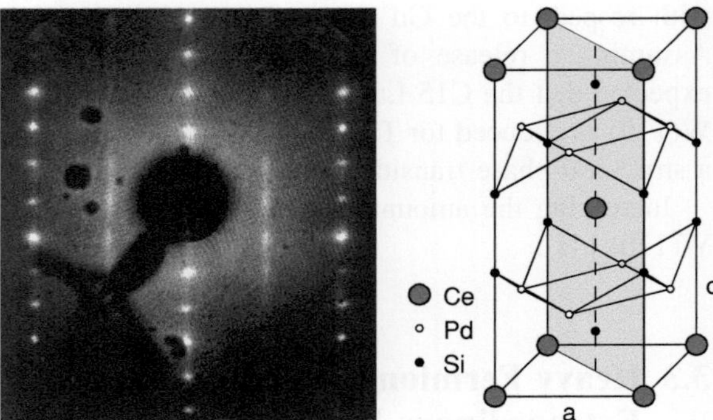

Fig. 3.25 *Left* LEED pattern of $CePd_2Si_2$ on W(110) at an energy of 58 eV. *Right* Crystal structure of $CePd_2Si_2$ for bulk material. The lattice constants are $a = 4.239$ Å and $c = 9.889$ Å. The *gray shaded area* shows the (110) plane being prepared in this investigation

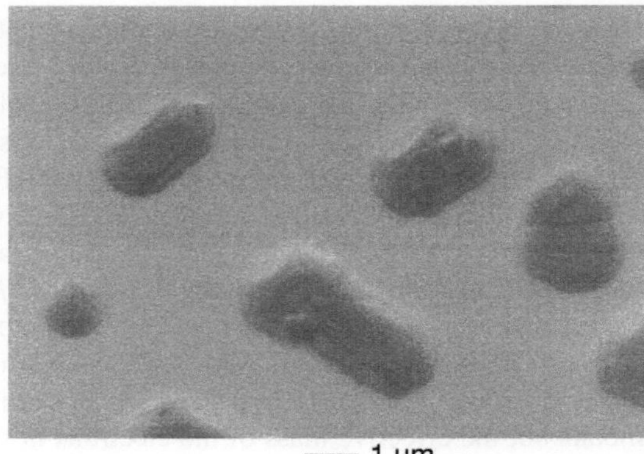

Fig. 3.26 The SEM image of $CeNi_2Ge_2$ of the layer shows crystalline islands which were identified by EDX to be Ni_2Ge. The total coverage of these islands corresponds to about 15% of the surface area. Reprinted from [47], Copyright (1997), with permission from Elsevier

$z = 0.379$. The measured distance between the rows of LEED spots reveals that the film is grown in [110] direction showing a (110) surface (see gray shaded area in the right part of Fig. 3.25).

The crystallization was additionally investigated using an electron microscope (see Fig. 3.26). The scanning electron microscopy (SEM) image together with an EDX (energy disperse X-ray detection) measurement revealed that on the smooth $CeNi_2Ge_2$ films small amounts of stable 3D crystalline islands of Ni_2Ge are formed which cover about 15% of the surface. Its appearance indicates a deficit of Ce on the surface most likely due to depletion during evaporation. Increasing the crucible temperature during evaporation appeared to reduce this depletion process.

3.3.2 Determination of Electronic Properties Using Resonant Photoemission

For the determination of electronic properties photoelectron spectroscopy with *variable* excitation energy was carried out because it is indispensable to

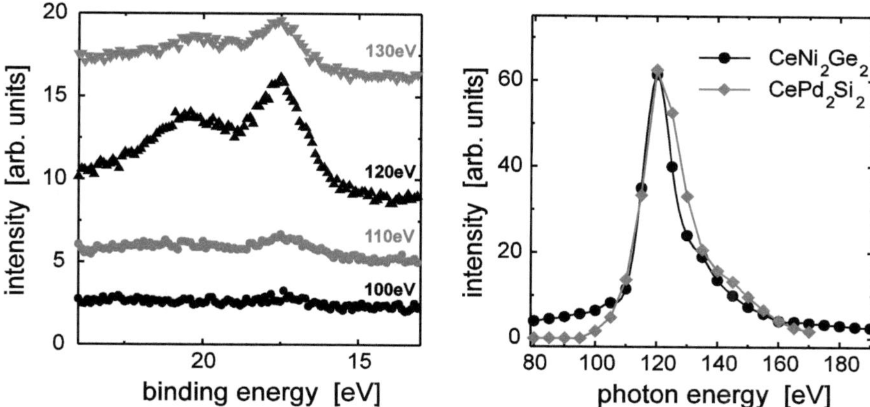

Fig. 3.27 a Energy distribution curves of CePd$_2$Si$_2$ showing the drastically enhanced emission of the Ce 5p states at an excitation energy of about 120 eV. **b** Peak intensities of the Ce 5p states as a function of photon energy for CePd$_2$Si$_2$ (*filled diamond*) and CeNi$_2$Ge$_2$ (*filled circle*). Reprinted with permission from [49]. Copyright (1998) by the American Physical Society

investigate the *orbital nature* as well as the *element character* for the different states being involved in the bonding of the intermetallic. For this purpose the technique of *resonant photoemission* was used.

The spectra of the energy region around 20 eV binding energy for excitation energies between 100 and 140 eV are shown in Fig. 3.27a for CePd$_2$Si$_2$. Obviously the emission of Ce 5$p_{1/2}$ and $p_{3/2}$ is drastically increased at about 120 eV. This behavior is directly correlated with the onset of the resonant enhancement of the 4$d \rightarrow$ 4f transition in the Ce compound. This process is described by 4d^{10} 4f^1 5p^6 + $hv \rightarrow$ 4d^9 4f^2 5p^6 with subsequent 4d^9 4f^2 5$p^6 \rightarrow$ 4d^{10} 4f^1 5p^5 + e$^-$ [50]. After a linear background subtraction the peak areas of the Ce 5p induced structures were determined as a function of the excitation energy. This result is presented in Fig. 3.27b for CePd$_2$Si$_2$ (filled diamond) and CeNi$_2$Ge$_2$ (filled circle). The giant enhancement of the cross section for a photon energy of about 120 eV is evident. The shape of these curves is in good agreement with theoretical calculations for metallic Ce [51].

The behavior of a giant resonance in the Ce 4f emission at $hv = 122$ eV [51] for photon induced excitations can be seen in Fig. 3.28 utilizing photon energies of 120 and 110 eV. This resonance was identified as a result of the transition 4d^{10} 4$f^1 \rightarrow$ 4d^9 4f^2 followed by an Auger decay process 4d^9 4$f^2 \rightarrow$ 4d^{10} 4f^0 + e$^-$, being a so-called "super-Coster–Kronig" transition [50]. Both spectra are normalized in such a way to give similar intensities in the energy region above 7 eV where no features are observed. The Pd, Si, Ni, and Ge derived photoemission intensities vary rather weakly with the photon energy. Therefore, by obtaining the difference spectrum, structures have to be Ce induced. The (120–110 eV) difference spectrum is shown in Fig. 3.28b for CePd$_2$Si$_2$. The appearance of the peak near E_F is not a result of the subtraction procedure but is already seen in the 120 eV spectrum.

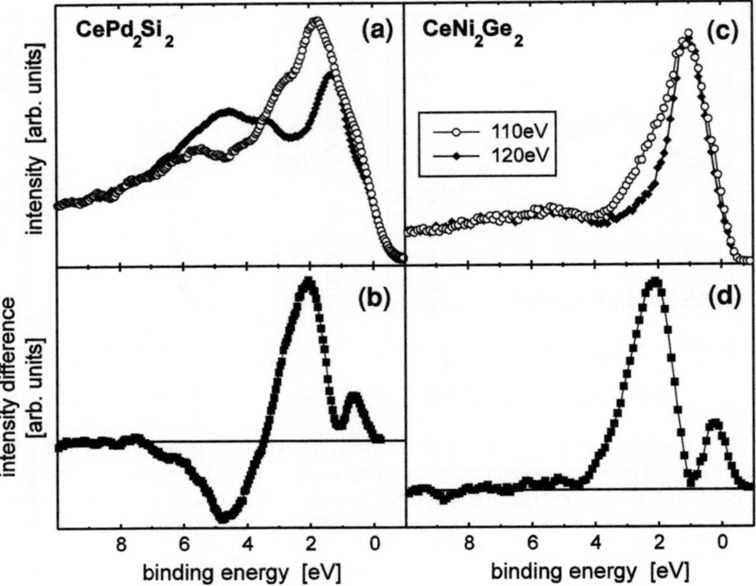

Fig. 3.28 Photoelectron spectra of **a** CePd$_2$Si$_2$ and **c** CeNi$_2$Ge$_2$. The spectra were taken at room temperature with an excitation energy of 110 and 120 eV. The corresponding difference spectra **b** and **d** point to Ce 4*f* derived emission at a binding energy of 2.0 and 0.5 eV due to a giant resonance at about 120 eV. Reprinted with permission from [49]. Copyright (1998) by the American Physical Society

A prior investigation of Parks et al. [52] determined both features at 2 and 0.5 eV binding energy to be Ce 4*f* derived. In contrast, their sample was bulk material which was fractured and the measurement was carried out in an angle integrated mode, thus only allowing the investigation of the density of states but not *k* resolved *E*(*k*) behavior. Their difference spectrum only showed both features near E_F; the one at 5 eV with opposite sign was not detected. Therefore, the existence of this feature demonstrates the possibilities of *angle resolved* investigations on high-quality and well-ordered surfaces. The corresponding investigation on CeNi$_2$Ge$_2$ (see Fig. 3.28c, d) does not show this structure. This directly points to a Pd induced character. A possible explanation may be the 4*d* character of Pd valence band electrons, in comparison to Ni 3*d*, being involved in a *d–f* hybridization.

The giant resonance in the Ce 4*f* emission facilitates to distinguish between the different hybridization characters of the bands being involved in the bonding of the intermetallic. This determination was carried out for CePd$_2$Si$_2$ using photon energies between 75 and 150 eV. The photoelectron intensities were determined for the three peaks in the energy region near the Fermi level at binding energies at 0.5, 2, and around 5 eV (cf. Fig. 3.28b). These values are shown in Fig. 3.29 (open circle: 0.5 eV, filled diamond: 2 eV, filled square: 5 eV). For excitation energies above 100 eV both peaks at low binding energy exhibit very low intensities and a

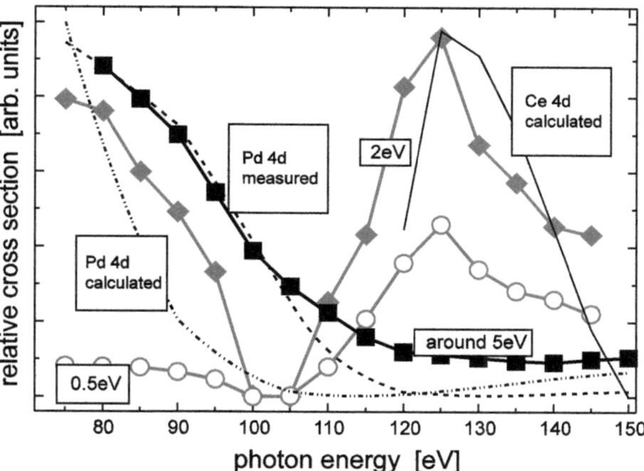

Fig. 3.29 Peak intensities of the three features near the Fermi energy for CePd$_2$Si$_2$ (*open circle, filled diamond, filled square*). Included are relative cross sections for Pd 4d (calculated: *dash-dotted* [53], measured: *dashed* [54]) and Ce 4d (calculated: *solid* [53]). Reprinted with permission from [49]. Copyright (1998) by the American Physical Society

giant enhancement at photon energies around 125 eV. This observation directly points to the f character of these electrons as already discussed above (cf. Fig. 3.28). The structure at a binding energy around 5 eV decreases slowly in intensity without any pronounced feature. Included in Fig. 3.29 are cross sections of Pd 4d states being experimentally determined [54] (dashed line) and theoretically calculated [53] (dash-dotted line), respectively. The shape of both curves is in good agreement to the intensity of the peak at 5 eV. This structure is therefore not caused by emission of Ce 4f states. Below an excitation energy of 100 eV the cross section of the peak at 0.5 eV increases not significantly to lower photon energies.

As already observed for Ce 4f states [55] this minimum is located around $hv = 100$ eV. In contrast, the structure at 2 eV increases rapidly to lower excitation energies. Below 100 eV the energy dependence of the cross section is therefore the same as for the 5 eV feature. This structure seems therefore to arise from emission of a band being caused by orbital mixing due to a not significantly weighted hybridization of Pd and Ce states. The theoretical calculated photoionization cross sections of Ce 4d states [53] are shown as a solid line. The shape of this curve with its maximum around $hv = 125$ eV demonstrates the f character of the photoemission peaks with binding energies of 0.5 eV and 2.0 eV.

Electron energy-loss spectroscopy at low excitation energies is a surface sensitive technique to study the electronic structure by exciting collective oscillations or electrons from occupied into unoccupied states. In metals with a high density of states arising from d electrons, the excitation of plasmon losses has a relatively low probability. Therefore, the spectra are dominated by interband or intraband transitions. In rare earth metals, excitations of the partially filled f shell are observed that are assigned to be dipole-forbidden 4f–4f transitions. These transitions are enhanced near the 4d–4f threshold [56].

The inset in Fig. 3.30 shows a typical electron energy-loss spectrum of CePd$_2$Si$_2$ taken at a primary energy of the incident electron beam of $E_p = 27$ eV. Two characteristic features are present in the spectrum: a pronounced peak at

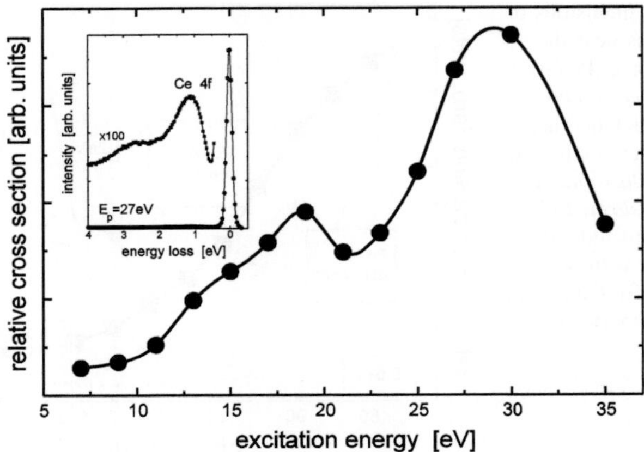

Fig. 3.30 Electron energy-loss spectrum of $CePd_2Si_2$ taken at room temperature with a primary energy of the incident electron beam of $E_p = 27$ eV; the prominent feature at a loss energy of 1.2 eV is caused by a dipole-forbidden $4f$–$4f$ transition of the Ce compound (see *inset*). The intensity of this feature is shown as a function of the electron excitation energy demonstrating a variation of the relative cross section. Reprinted with permission from [49]. Copyright (1998) by the American Physical Society

1.2 eV and a weak one at an energy loss between 2 and 3.5 eV. In comparison with band structure calculations [57, 58] and ultraviolet photoelectron spectroscopy (UPS), the broad feature at about 2.5 eV can be assigned to be an interband transition. The loss at 1.2 eV represents a dipole forbidden $4f$–$4f$ transition of the Ce compound. This value corresponds to the energy difference of the occupied and unoccupied $4f$ levels and gives evidence for a $4f$ binding energy of below 1.2 eV. The intensity of this feature is shown as a function of the electron excitation energy demonstrating a variation of the relative cross section. The relationship between excitation of the Ce $5p$ state and the $4f$–$4f$ transition is obvious due to the maximum of the relative cross section around 18 eV. The second maximum at 28 eV points to an additional enhancement of the cross section in the dipole forbidden f–f transition.

The final state in this excitation spectroscopy is a neutral excited state because no electron has been emitted. Therefore, the determination of electronic properties of $4f$ states will be useful for a *direct* comparison with band structure calculations because in the photoemission process from a localized state an ionic final state is created due to the emission of the photoelectron. Thus the final state energy can be strongly shifted as compared with the neutral complex. As already shown in UPS and EELS investigations on rare earth metals (e.g., on Gd [59]), the binding energies determined by these techniques differ up to about 4 eV.

To summarize the last part, it was demonstrated that atomically clean and well-ordered surfaces of ternary heavy fermion intermetallics were prepared allowing the application of electron spectroscopic methods being sensitive to surface properties. The structural and electronic properties of $CePd_2Si_2$ and $CeNi_2Ge_2$

were investigated using the techniques of LEED, MEED (medium-energy electron diffraction), AES [47], UPS [46], and EELS. Characteristic MEED oscillations demonstrate a layer-by-layer growth [45]. EELS allowed to investigate the binding energy of the Ce $4f$ states. The giant resonance in the Ce $4f$ emission facilitated the determination of $4f$ derived features in the photoelectron spectra [49].

References

1. D. Weller, S. Alvarado, W. Gudat, K. Schröder, M. Campagna, Phys. Rev. Lett. **54**, 1555 (1985)
2. M. Farle, W. Lewis, K. Baberschke, Appl. Phys. Lett. **62**, 2728 (1993)
3. H. Tang, D. Weller, T. Walker, J. Scott, C. Chappert, H. Hopster, A. Pang, D. Dessau, D. Pappas, Phys. Rev. Lett. **71**, 444 (1993)
4. E. Weschke, C. Schüssler-Langeheine, R. Meier, A. Fedorov, K. Starke, F. Hübinger, G. Kaindl, Phys. Rev. Lett. **77**, 3415 (1996)
5. D. Li, J. Pearson, S. Bader, D. McIlroy, C. Waldfried, P. Dowben, J. Appl. Phys. **79**, 5838 (1996)
6. D. Li, C. Hutchings, P. Dowben, R. Wu, C. Hwang, M. Onellion, A. Andrews, J. Erskine, J. Appl. Phys. **70**, 6565 (1991)
7. A. Fedorov, K. Starke, G. Kaindl, Phys. Rev. B **50**, 2739 (1994)
8. M. Farle, W. Lewis, J. Appl. Phys. **75**, 5604 (1994)
9. M. Farle, K. Baberschke, U. Stetter, A. Aspelmeier, F. Gerhardter, Phys. Rev. B **47**, 11571 (1993)
10. J. Suits, D. Rugar, C. Lin, J. Appl. Phys. **64**, 252 (1988)
11. T. Morishita, Y. Togami, K. Tsushima, J. Phys. Soc. Jpn. **54**, 37 (1985)
12. C. Wang, R. Osgood, R. White, B. Clemens, in *Magnetic Ultrathin Films, Multilayers and Surfaces*, ed. by E. Marinero, B. Heinrich, W. Egelhoff, A. Fert, H. Fujimori, G. Güntherodt, R. White (1995) (Mater. Res. Soc. Symp. Proc., San Francisco, 1995)
13. F. Robaut, P. Milkulik, N. Cherief, O. McGrath, D. Givord, T. Baumbach, J. Veuillen, J. Appl. Phys. **78**, 997 (1995)
14. V. Oderno, C. Dufour, K. Dumesnil, P. Mangin, G. Marchal, J. Cryst. Growth **165**, 175 (1996)
15. M. Huth, C. Flynn, Phys. Rev. B **58**, 11526 (1998)
16. F. Steglich, J. Aarts, C. Bredl, W. Lieke, D. Meschede, W. Franz, H. Schäfer, Phys. Rev. Lett. **43**, 1892 (1979)
17. C. Bredl, S. Horn, F. Steglich, B. Lüthi, R. Martin, Phys. Rev. Lett. **52**, 1982 (1984)
18. H. Aoki, S. Uji, A. Albessard, Y. Ōnuki, Phys. Rev. Lett. **71**, 2110 (1993)
19. C. Laubschat, E. Weschke, C. Holtz, M. Domke, O. Strebel, G. Kaindl, Phys. Rev. Lett. **65**, 1639 (1990)
20. E. Weschke, C. Laubschat, R. Ecker, A. Höhr, M. Domke, G. Kaindl, L. Severin, B. Johansson, Phys. Rev. Lett. **69**, 1792 (1992)
21. R. Parks, S. Raaen, M. den Boer, Y.S. Chang, G. Williams, Phys. Rev. Lett. **52**, 2176 (1984)
22. D. Weller, S. Alvarado, J. Appl. Phys. **59**, 2908 (1986)
23. J. Kołaczkiewicz, E. Bauer, Surf. Sci. **175**, 487 (1986)
24. E. Tober, R. Ynzunza, C. Westphal, C. Fadley, Phys. Rev. B **53**, 5444 (1996)
25. R. Pascal, C. Zarnitz, H. Tödter, M. Bode, M. Getzlaff, R. Wiesendanger, Appl. Phys. A **66**, S1121 (1998)
26. R. Pascal, C. Zarnitz, M. Bode, R. Wiesendanger, Phys. Rev. B **56**, 3636 (1997)
27. R. Pascal, C. Zarnitz, M. Bode, R. Wiesendanger, Surf. Sci. **385**, L990 (1997)
28. R. Szukiewicz, J. Kołaczkiewicz, I. Yakovkin, Surf. Sci. **602**, 2610 (2008)

29. M. Wiejak, M. Jankowski, I. Yakovkin, J. Kołaczkiewicz, Appl. Surf. Sci. **256**, 4834 (2010)
30. S. Nepijko, M. Getzlaff, R. Pascal, C. Zarnitz, M. Bode, R. Wiesendanger, Surf. Sci. **466**, 89 (2000)
31. A. Aspelmeier, F. Gerhardter, K. Baberschke, J. Magn. Magn. Mater. **132**, 22 (1994)
32. D. Li, P. Dowben, J. Ortega, F. Himpsel, Phys. Rev. B **49**, 7734 (1994)
33. M. Bode, M. Getzlaff, R. Pascal, S. Heinze, R. Wiesendanger, in *Magnetism and Electronic Correlations in Local-Moment Systems: Rare Earth Elements and Compounds*, ed. by M. Donath, P. Dowben, W. Nolting (World Scientific, Singapore, 1998)
34. R. Pascal, C. Zarnitz, M. Bode, M. Getzlaff, R. Wiesendanger, Appl. Phys. A **65**, 603 (1997)
35. M. Wuttig, X. Liu, *Ultrathin Metal Films* (Springer, Berlin, 2004)
36. H. Bethge, D. Heuer, C. Jensen, K. Reshöft, U. Köhler, Surf. Sci. **331–333**, 878 (1995)
37. M. Bode, R. Pascal, M. Dreyer, R. Wiesendanger, Phys. Rev. B **54**, 8385 (1996)
38. R. Pascal, M. Getzlaff, H. Tödter, M. Bode, R. Wiesendanger, Phys. Rev. B **60**, 16109 (1999)
39. M. Getzlaff, R. Pascal, H. Tödter, M. Bode, R. Wiesendanger, Appl. Surf. Sci. **142**, 543 (1999)
40. M. Getzlaff, R. Pascal, H. Tödter, M. Bode, R. Wiesendanger, Surf. Rev. Lett. **6**, 741 (1999)
41. K. Gschneidner, L. Eyring (eds.), *Handbook of the Physics and Chemistry of Rare Earths*, vol. 2 (North Holland, Amsterdam, 1979)
42. M. Getzlaff, R. Pascal, R. Wiesendanger, Surf. Sci. **566-568**, 236 (2004)
43. K. Andres, J. Graeben, H. Ott, Phys. Rev. Lett. **35**, 1779 (1979)
44. G. Stewart, Z. Fisk, J. Willis, J. Smith, Phys. Rev. Lett. **52**, 679 (1984)
45. B. Schmied, M. Wilhelm, U. Kübler, M. Getzlaff, G. Fecher, G. Schönhense, Fresenius J. Anal. Chem. **358**, 141 (1997)
46. B. Schmied, M. Wilhelm, U. Kübler, M. Getzlaff, G. Fecher, G. Schönhense, Physica B **230–232**, 290 (1997)
47. B. Schmied, M. Wilhelm, U. Kübler, M. Getzlaff, G. Fecher, G. Schönhense, Surf. Sci. **377–379**, 251 (1997)
48. A. Palenzona, S. Cirafici, F. Canepa, J. Less–Common Met. **135**, 185 (1987)
49. M. Getzlaff, B. Schmied, M. Wilhelm, U. Kübler, G. Fecher, J. Bansmann, L. Lu, G. Schönhense, Phys. Rev. B **58**, 9670 (1998)
50. W. Lenth, F. Lutz, J. Barth, G. Kalkoffen, C. Kunz, Phys. Rev. Lett. **41**, 1185 (1978)
51. A. Zangwill, P. Soven, Phys. Rev. Lett. **45**, 204 (1980)
52. R. Parks, B. Reihl, N. Mårtensson, F. Steglich, Phys. Rev. B **27**, 6052 (1983)
53. J. Yeh, I. Lindau, Atomic Data Nucl. Data Tables **32**, 1 (1985)
54. G. Rossi, I. Lindau, L. Braicovich, I. Abbati, Phys. Rev. B **28**, 3031 (1983)
55. A. Franciosi, J. Weaver, N. Mårtensson, M. Croft, Phys. Rev. B **24**, 3651 (1981)
56. S. Modesti, G. Paolucci, E. Tosatti, Phys. Rev. Lett. **55**, 2995 (1985)
57. J. Sticht, N. Ambrumenil, J. Kübler, Z. Phys. B **65**, 149 (1986)
58. E. Runge, R. Albers, N. Christensen, G. Zwicknagel, Phys. Rev. B **51**, 10375 (1995)
59. D. Weller, S. Alvarado, Z. Phys. B **58**, 261 (1985)

Chapter 4
Influence of Adsorbates

The *bare* surfaces of rare earth metal systems were described above with respect to their structural and electronic behavior. As already mentioned occasionally in the previous chapter, adsorbates, e.g. from the residual gas, alter the electronic properties of the clean surfaces due to the hybridization of adsorbate atom and substrate states (for example, the Gd surface state can only exist if the surface is atomically clean). The strong influence of adsorbate atoms on the surface due to this chemisorption process does not only induce variations in the electronic properties—this will be discussed in Sects. 4.1 and 4.4 for hydrogen and oxygen being adsorbed on Gd(0001), respectively—but may also lead to distinct structural changes (e.g., surface reconstructions) which will exemplarily be shown in Sect. 4.2 for hydrogen being dissolved in gadolinium thin films. Using the distinct surface sensitivity of both techniques—scanning tunneling microscopy (STM) and photoelectron spectroscopy—it will be shown in Sect. 4.3, by describing the coadsorption of hydrogen and CO on gadolinium surfaces, that their combination allows a significantly more detailed analysis.

4.1 Unusual Adsorption Characteristics for Hydrogen on Gd Surfaces

In this part the electronic properties of chemisorbed hydrogen on Gd(0001) will be discussed. For the photoemission investigation, smooth Gd(0001) films were prepared and subsequently exposed to hydrogen. It was already shown in Sect. 3.1 (Chap. 3) that the Gd surface state exists both on smooth films and multilayer Gd islands. Therefore, multilayer island films were chosen for the STM investigations to provide topographical contrast in order to distinguish between sample states and tip induced artifacts as well as to determine the influence of different island heights. The detailed preparation procedure was already described in Sect. 3.1 (Chap. 3).

M. Getzlaff, *Surface Magnetism*, Springer Tracts in Modern Physics, 240,
DOI: 10.1007/978-3-642-14189-8_4, © Springer-Verlag Berlin Heidelberg 2010

4.1.1 The Hydrogen Induced State on Gd(0001)

The determination of the differential cross section plays an important role describing the photoemission process. For a fixed spatial arrangement the angular distribution of photoelectrons can be deduced if the dipole matrix elements and the phase shifts of the outgoing photoelectron wave are known. One example for the determination of these values is the Circular Dichroism in the Angular Distribution of photoelectrons (CDAD) [1]. Using this experimental technique the photoelectron intensity as a function of detection angle is compared for excitation with left and right circularly polarized light. The important parameters describing the intensity difference are the dipole matrix elements and the relative phase shifts. The capability of this technique will be demonstrated for oxygen adsorbed on (0001) surfaces of rare earth metals (see Sect. 4.4).

This type of investigation is usually carried out with a *fixed polarization* of the incoming light and a variation of the emission angle. A disadvantage of this experimental arrangement is given by the limited range of the detection angle being further reduced due to the refraction of the escaping photoelectrons at the electrostatic surface barrier being explained below. Measurements at a *fixed detection angle* with a rotating electrical field vector enables to avoid this restriction due to the possibility of using the whole angle range of 180° giving a significantly more accurate set of data.

Additionally, information on quantum mechanical quantities are usually obtained with a more or less larger experimental or theoretical effort. Being in contrast, the theoretical description of the latter type of angle resolving photo-emission experiments as well as the experimental procedure are relatively simple. This method is suitable to be carried out in a laboratory because it does not need a sophisticated experimental setup as synchrotron radiation sources. It may therefore play an important role for a proceeding understanding of the photoemission process.

In the following it will be reported on a straightforward method to provide information about the dipole matrix elements and phase shifts being essential for the theoretical description of the photoemission process in a relatively simple way and with a pronounced accuracy [2]. This can be achieved by means of photo-electron spectroscopy with linearly polarized light using the ability of a *continuous rotation of the electric field vector*. The method is exemplarily demonstrated at the system hydrogen on Gd(0001)/W(110) which possesses a pronounced d_{z^2}-like adsorbate induced state.

Photoelectrons were produced using linearly polarized vacuum ultraviolet (VUV) radiation from a discharge lamp (Ne I resonance line: $hv = 16.85$ eV) with a triple reflection polarizer. The angle of the incoming photon beam was $\theta_{ph} = 45°$ with respect to the surface normal. The experimental setup is schematically shown in Fig. 4.1.

Photoelectron spectra from the clean Gd(0001) surface and after hydrogen exposures are presented in Fig. 4.2. The spectra were taken in normal emission at room temperature. The sharp feature near the Fermi edge is due to the Gd surface

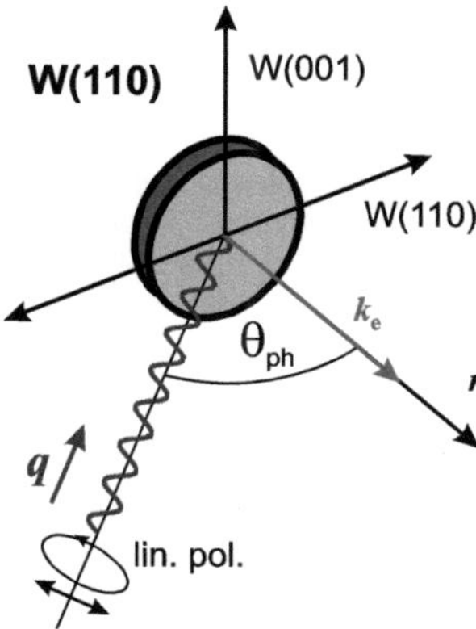

Fig. 4.1 Schematic view of the experimental setup. The **E** vector of the incoming linearly polarized radiation can be rotated as indicated by the *circle*

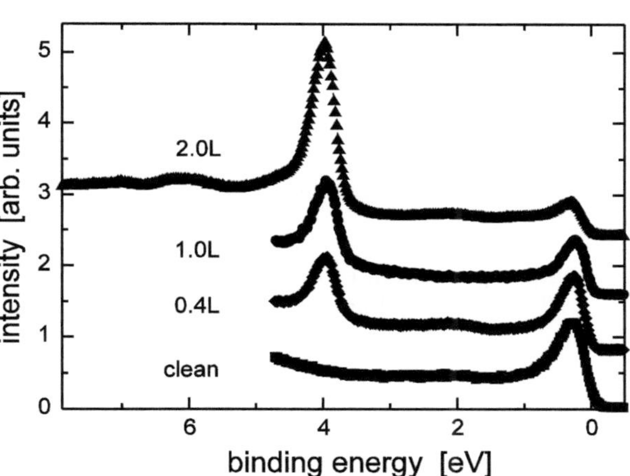

Fig. 4.2 Photoemission spectra at normal emission for different hydrogen exposures on Gd(0001)/W(110). The photon energy is 16.85 eV. Reprinted with permission from [3]. Copyright (1999) by the American Physical Society

state and is only observed on contamination-free high-quality hcp(0001) surfaces. After hydrogen exposure a pronounced feature at a binding energy of about 4 eV appears. The energy width, defined as full width at half maximum (FWHM), of this hydrogen induced state amounts to 0.4 eV and is therefore similar to that of the surface state. The weak feature at about 2 eV is caused by Gd Δ_2 bulk bands [4]. Dosing additional hydrogen suppresses the Gd surface state as previously demonstrated by Li et al. [5].

In Fig. 4.3 photoelectron spectra of the H induced state are presented for different angles α of the **E** field vector of the incoming linearly polarized photon beam with respect to the plane of incidence. $\alpha = 90°$ denotes the case for s polarized light, and it is this angle where the intensity of the H induced state nearly vanishes whereas the intensity is significantly enhanced for more p polarized light.

The photoelectron intensity I (with the detector being at infinity) for atomic orbitals which are excited by linearly polarized light can be calculated via the differential cross section $d\sigma/d\Omega$ [6]:

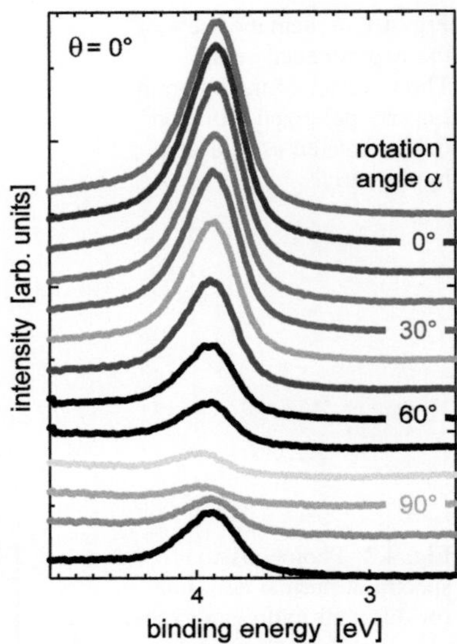

Fig. 4.3 Photoelectron spectra at $\theta = 0°$ (normal emission) for the hydrogen induced structure as a function of the rotation angle α of the linear polarizer. Reprinted from [7], Copyright (1998), with permission from Elsevier

$$I = \frac{d\sigma}{d\Omega} = \frac{4\pi}{3}\alpha a_0^2 h\nu \left| \left\langle \phi_{E_{\mathrm{kin}},\mathbf{k}} | \boldsymbol{\varepsilon} \cdot \mathbf{r} | \phi_{n\ell x} \right\rangle \right|^2 \tag{4.1}$$

with α being the fine structure constant, a_0 the Bohr radius, $\mathbf{k}$ the wave vector of the outgoing electron, $\boldsymbol{\varepsilon}$ the polarization vector, and $\mathbf{r}$ the position vector at angles θ, ϕ. The initial real atomic orbital is given by

$$\phi_{n\ell x} = R_{n\ell}(r) \sum_m n(m) Y_{\ell m}(\theta, \phi) \tag{4.2}$$

with $Y_{\ell m}(\theta, \phi)$ being a spherical harmonic and $n(m)$ the coefficients needed to form real orbitals such that ℓx is s, p_z, etc. The final state is written [8] as a partial wave expansion

$$\phi_{E_{\mathrm{kin}},\mathbf{k}} = 4\pi \sum_{\ell' m'} i^{\ell'} e^{-i\delta_{\ell'}} Y_{\ell',m'}^*(\theta_k, \phi_k) Y_{\ell',m'}(\theta, \phi) R_{E_{\mathrm{kin}},\ell'} \tag{4.3}$$

Using these Eqs. 4.1–4.3 and dipole selection rules $\Delta\ell = \pm 1$ and $\Delta m = \pm 1, 0$ the dipole matrix element can be written as

$$\begin{aligned}
\left\langle \phi_{E_{\mathrm{kin}},\mathbf{k}} | \boldsymbol{\varepsilon} \cdot \mathbf{r} | \phi_{n\ell x} \right\rangle = {}& \sqrt{8\pi} \sum_{\ell' m} n(m)(-i)^{\ell'} e^{i\delta_{\ell'}} \\
& \times R_{\ell'}(E_{kin})(\varepsilon_x[-Y_{\ell',m+1}(\theta_k, \phi_k)C^1(\ell', m+1, \ell, m) \\
& + Y_{\ell',m-1}(\theta_k, \phi_k)C^1(\ell', m-1, \ell, m)] \\
& + i\varepsilon_y[Y_{\ell',m+1}(\theta_k, \phi_k)C^1(\ell', m+1, \ell, m) \\
& + Y_{\ell',m-1}(\theta_k, \phi_k)C^1(\ell', m-1, \ell, m)] \\
& + \sqrt{2}\varepsilon_z Y_{\ell',m}(\theta_k, \phi_k)C^1(\ell', m, \ell, m))
\end{aligned} \tag{4.4}$$

with the Gaunt coefficients $C^1(\ell', m', \ell, m)$ [9].

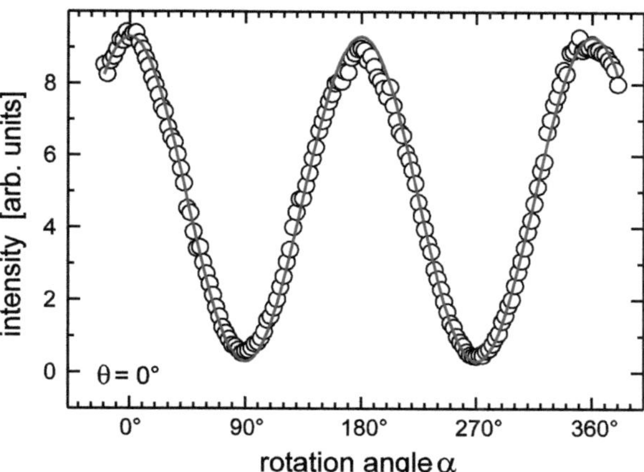

Fig. 4.4 Photoemission intensities of the maximum for the hydrogen induced feature at a binding energy of about 4 eV as a function of the rotation angle α of the incoming linearly polarized radiation; the solid line represents a fit with a $\cos^2$ function. The spectra were taken in normal emission. Reprinted from [7], Copyright (1998), with permission from Elsevier

The photoelectron intensity of the hydrogen induced state was determined (cf. Fig. 4.3) as a function of the vector potential of the incoming radiation and is shown in Fig. 4.4 (open circles). The electric field vector of the light was rotated by the angle α in the plane perpendicular to the direction of propagation. For $\alpha = 0°$, the light is highly p polarized, whereas for $\alpha = 90°$ it is s polarized (cf. Fig. 4.1). The intensity exhibits maxima at $0°$ and $180°$, respectively. A d_{z^2}-like orbital symmetry of this state was previously deduced by Li et al. [5] from the observation that the constant initial state spectra were the same for both the hydrogen induced state and the Gd surface state. The dipole matrix element for such a d_{z^2} initial state can be calculated to be

$$
\langle \phi_{E_{\text{kin}},\mathbf{k}} | \boldsymbol{\varepsilon} \cdot \mathbf{r} | d_{z^2} \rangle = \sqrt{8\pi}\left[e^{i\delta_p} R_p \left(\sqrt{1/10}\,\sin\theta (\varepsilon_x \cos\phi + \varepsilon_y \sin\phi) - \varepsilon_z \sqrt{2/5}\,\cos\theta \right) \right.
$$
$$
+ e^{i\delta_f} R_f \sqrt{9/40}(\sin\theta (5\cos^2\theta - 1)(\varepsilon_x \cos\phi + \varepsilon_y \sin\phi)
$$
$$
\left. + \varepsilon_z (5\cos^3\theta - 3\cos\theta)) \right]
\tag{4.5}
$$

In normal emission ($\theta = 0$) this expression is determined by ε_z being proportional to $\cos\alpha$. The intensity using a rotating electric field vector is therefore proportional to $\cos^2\alpha$ (see Eq. 4.1). This behavior is demonstrated by the solid line in Fig. 4.4 presenting a fit with a $\cos^2$ function.

Dispersion effects of the parallel momentum component $k_\parallel$ from the outgoing electron was investigated by rotating the sample with the radiation source and analyzer kept fixed in space. The photoelectron spectra for angles between $\theta = 0°$ and $60°$ are shown in Fig. 4.5. Whereas in normal emission the spectrum is dominated by the hydrogen induced structure at about 4 eV binding energy, this feature looses intensity with increasing detection angle and shifts in energy directly pointing to a distinct dispersion. For higher angles additional structures at binding energies of 1 and 5.5 eV appear. Figure 4.6 shows the experimentally determined band dispersion along the $\bar{\Gamma}$–$\bar{M}$-direction in the Gd surface Brillouin

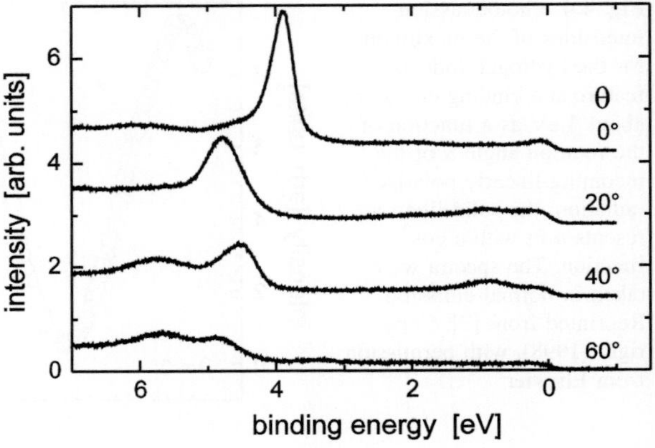

Fig. 4.5 Photoelectron spectra for different detection angles θ. Variations in binding energy, pointing to a dispersion, as well as in intensity are present. Reprinted from [7], Copyright (1998), with permission from Elsevier

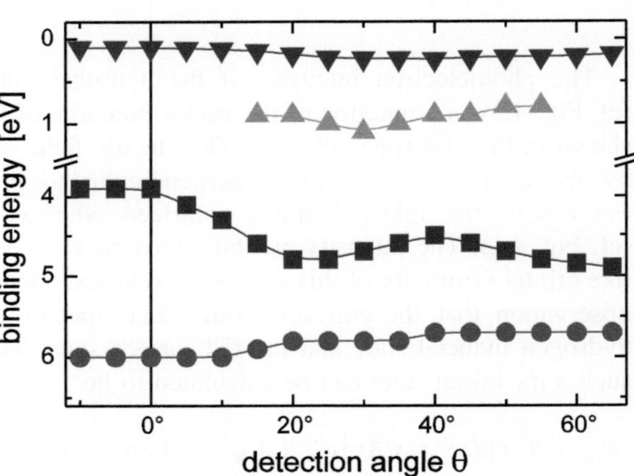

Fig. 4.6 Two-dimensional band dispersion along the $\bar{\Gamma}-\bar{M}$ direction of the Gd surface Brillouin zone. Reprinted from [7], Copyright (1998), with permission from Elsevier

zone. The highly localized Gd surface state near E_F exhibits nearly no dispersion. In contrast, the H induced states possess different binding energies when varying the photoelectron detection angle θ. The $k_\parallel$ dispersion suggests an overlap of the hydrogen wave functions within the overlayer or hybridization with the Gd bulk bands. Since the electronic properties of this hydrogen induced state are different from those of the Gd surface state it is clearly not a surface state.

The dipole matrix element contains the radial parts R_p and R_f as well as phase shifts δ_p and δ_f (see Eq. 4.5). In order to obtain information about these properties the photoelectron intensities were determined at a fixed detection angle $\theta = 45°$ as a function of the rotation of the **E**-field vector. The spectra in Fig. 4.7 are shown for particular values of α which the maximum and minimum intensities are reached at for the peak 2 at 4.7 eV (4.0 eV in normal emission) with $\alpha = 170°$ and $80°$ as well as for the feature 1 at 1 eV and structure 3 at 6 eV with $\alpha = 140°$ and $50°$, respectively. The intensity values are summarized in Fig. 4.8 (filled diamond: Peak 1, open circle: Peak 2, filled square: Peak 3). The curves for peak 1 and 3 exhibit the same shape which may be caused by emission from orbitals with the

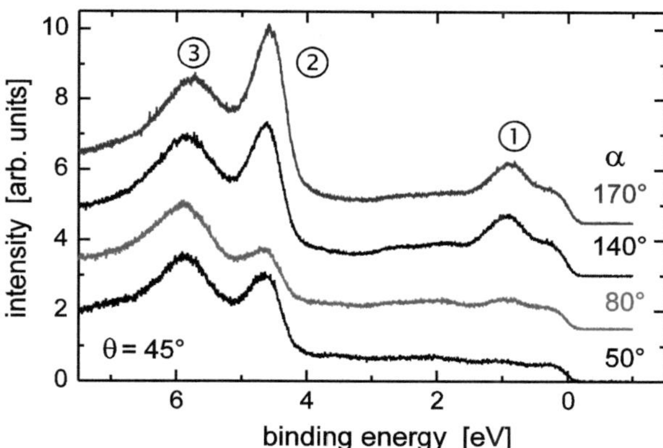

Fig. 4.7 Photoemission intensities at a fixed detection angle $\theta = 45°$ for particular values of the rotation angle α (see text). Reprinted from [7], Copyright (1998), with permission from Elsevier

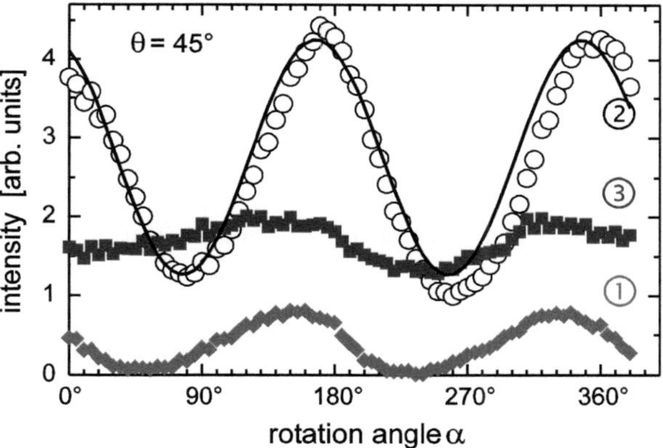

Fig. 4.8 Photoemission intensities of the maximum for the three peaks labeled in Fig. 4.7. The shape of *curves 1* and *3* is the same one but different to the one for curve *2*. The *solid line* represents a fit as discussed in the text. Reprinted from [7], Copyright (1998), with permission from Elsevier

same symmetry but are different with respect to the d_{z^2}-like state (peak 2). Equation 4.4 can be expressed in terms of quantities $X_{\ell\pm1,x}$ [6] as follows:

$$\left\langle \phi_{E_{\mathrm{kin}},\mathbf{k}} | \boldsymbol{\varepsilon} \cdot \mathbf{r} | \phi_{n\ell x} \right\rangle = (-i)^{\ell-1}\left(e^{i\delta_{\ell-1}} X_{\ell-1,x} + e^{i\delta_{\ell+1}} X_{\ell+1,x} \right) \tag{4.6}$$

For this experimental geometry these X-functions are given by

$$X_{\ell-1,x} = R\left(\sin\theta\left(\sin\alpha + \sqrt{1/2}\cos\alpha \right) - 2\cos\theta\left(\sin\alpha - \sqrt{1/2}\cos\alpha \right) \right) \tag{4.7}$$

$$\begin{aligned}
X_{\ell+1,x} = 1.5\Big(&\sin\theta(5\cos^2\theta - 1)(\sin\alpha + \sqrt{1/2}\cos\alpha) \\
&+ \cos\theta(5\cos^2\theta - 3)(\sin\alpha - \sqrt{1/2}\cos\alpha)\Big)
\end{aligned} \tag{4.8}$$

using the simplification $R_f = 1$ and $R = R_p/R_f$. The photoelectron intensity can then be written as

$$I \propto X_{\ell+1,x}^2 + X_{\ell-1,x}^2 + 2X_{\ell+1,x}X_{\ell-1,x}\cos(\delta_{\ell+1} - \delta_{\ell-1}) \tag{4.9}$$

However, the angle θ in these equations is not the detected one as can be seen by the following consideration.

In off-normal measurements ($\theta \neq 0$) the detection angle θ with the detector at infinity corresponds to a smaller one in or near the crystal. This effect is due to the refraction of an outgoing electron wave at the electrostatic surface barrier. When surmounting this barrier, the momentum parallel to the surface remains constant, the one perpendicular becomes reduced. This phenomenon results in an increased emission angle (with respect to the surface normal) outside the surface region. The inner, i.e. true, emission angle θ^{in} can be deduced from the outer, i.e. observed, angle $\theta^{\mathrm{out}} = \theta$ using the relation [10, 11]

$$\sqrt{E_{\mathrm{kin}} + V_0} \cdot \sin \theta^{\mathrm{in}} = \sqrt{E_{\mathrm{kin}}} \cdot \sin \theta^{\mathrm{out}} \tag{4.10}$$

with E_{kin} being the kinetic energy of the photoelectron and V_0 the inner potential being the difference in kinetic energy inside and outside the surface barrier. The binding energy of the hydrogen induced state corresponds to a kinetic energy of about 10 eV. Assuming a typical value for the inner potential of $V_0 = -10$ eV the maximum value for the detection angle $\theta^{\mathrm{out}} = 90°$ corresponds to an inner angle $\theta^{\mathrm{in}} = 45°$; i.e. the "observable" angle range is limited to this value of 45°. The detection angle $\theta = 45°$ (cf. Fig. 4.7) therefore corresponds to an inner angle $\theta^{\mathrm{in}} = 30°$ which must be used in Eqs. 4.7 and 4.8. A least-squares fit for the d_{z^2}-like hydrogen induced structure (curve 2 in Fig. 4.8) results in $R = 2.4 \pm 0.3$ and $\delta_f - \delta_p = 310° \pm 10°$. In order to verify these values the intensity of the same feature was determined as a function of the detection angle (cf. Fig. 4.5). These values are presented in Fig. 4.9. Calculating the X-function for this experimental geometry and inserting $R = 2.4$ and $\delta_f - \delta_p = 310°$ into the equation results in the function which is shown as a solid line in Fig. 4.9. There is good agreement within the experimental error corroborating the findings for the ratio of the radial matrix elements and the relative phase shift. It should be mentioned that these considerations are carried out within a pure atomic model being an approximate description although this d_{z^2}-like hydrogen state is largely atomic-like.

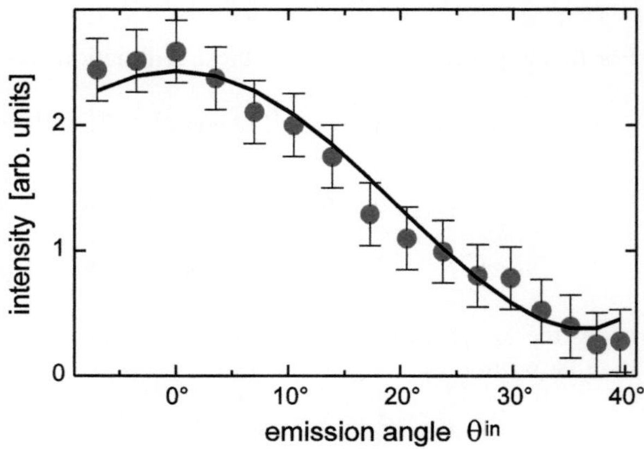

Fig. 4.9 Photoemission intensities as a function of the detection angle θ. The angle scale is given by the emission angle θ^{in} inside the surface barrier which can be deduced from the detection angle taking into account the inner potential (see Eq. 4.10). Reprinted from [7], Copyright (1998), with permission from Elsevier

4.1.2 Spatially Resolved Structural and Electronic Properties

As shown in the photoelectron spectra (see Fig. 4.2), the hydrogen induced state exhibits a binding energy of about 4 eV and is therefore nearly unaccessible to STM investigations. It is nevertheless possible to resolve the spatial distribution of hydrogen since the suppression of the Gd surface state leads to a drastic reduction of the differential conductance at low bias voltages.

This behavior is illustrated in Fig. 4.10 which shows the tunneling dI/dU spectra measured on Gd(0001) island surfaces (solid curve), on the first Gd monolayer between the islands (dash-dotted line), and on areas which have been modified by hydrogen adsorption (dashed line). The suppression of the gadolinium surface state is obvious. The strong decrease in the differential conductance which is basically proportional to the local density of states (LDOS) at the surface demonstrates the strong lateral localization of the surface state; bulk states do not show such pronounced differences. The tunneling spectra reveal that for the Gd monolayer only one peak at about 0.3 V is present (see Fig. 4.10) which cannot be attributed to the surface state. This can be deduced from the observation that no changes in the tunneling spectra occur after hydrogen adsorption; a surface state, in contrast, is very sensitive to adsorbates (cf. Fig. 4.2). Additionally, the determination of the energies of both spin counterparts from the surface state as a function of temperature shows a shift towards the Fermi level; the feature of the monolayer, however, remains fixed in energy [12]. This will be discussed in more detail in Sect. 5.2.1 (Chap. 5). Summarizing, hydrogen adsorption does not significantly modify the electronic structure of the monolayer.

In Fig. 4.11a and b, constant-current topographs of Gd islands on W(110) after different hydrogen exposures are shown. Between the high Gd islands which exhibit a (0001) surface, a Gd monolayer covers the tungsten substrate. In addition, two and three monolayer high islands are visible. The steps of the tungsten substrate even remain visible on the high Gd islands. The lines arise from lattice dislocations through the whole island due to different lattice constants in perpendicular direction.

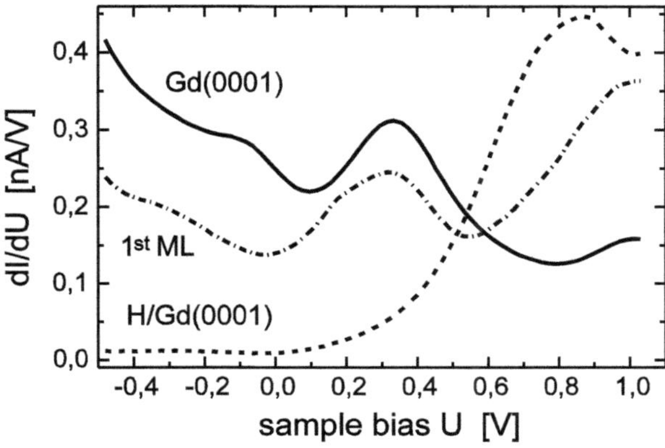

Fig. 4.10 Tunneling *dI/dU* spectra (stabilization parameters: $U = 1.0$ V, $I = 0.3$ nA) measured on Gd island surfaces (*solid*), on the first Gd monolayer between the islands (*dash-dotted*), and on areas which have been affected by hydrogen adsorption (*dashed*). Reprinted from [7], Copyright (1998), with permission from Elsevier

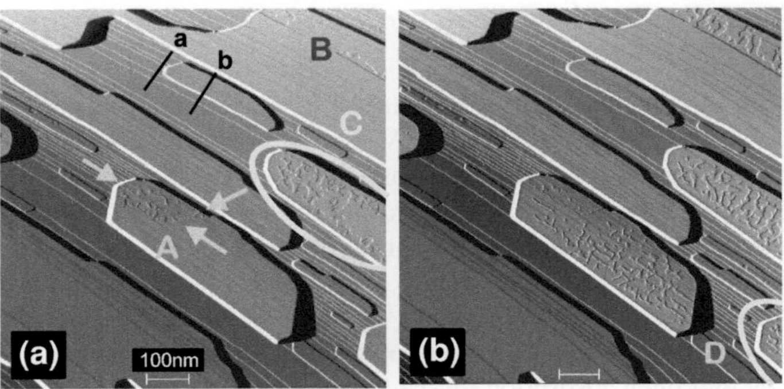

Fig. 4.11 Constant-current topograph ($U = -0.3$ V, $I = 0.03$ nA) of Gd islands on W(110). Between high Gd islands with a (0001) surface, a Gd monolayer covers the tungsten substrate. Additionally, two and three monolayer high islands are visible. **a** To this system, 0.2 L hydrogen was dosed. **b** The same area as in (**a**) but with an additional dosage of 0.8 L resulting in a total amount of 1.0 L hydrogen. Reprinted with permission from [3]. Copyright (1999) by the American Physical Society

Figure 4.11a shows this system after an exposure of 0.2 L hydrogen. The gadolinium monolayer and the low islands remain nearly unaffected. The monolayer of Gd on W(110) possesses a strained hcp structure with a dilatation of about 8% [13]. Two different kinds of behavior can be observed on top of the high islands. Some of them show a smooth surface which is undisturbed by hydrogen. Others present a lagoon-like appearance of alterations. In the regions with a "collapsed" appearance the Gadolinium surface state is suppressed whereas in the unaltered regions it still exists. Due to the reduction in the differential conductance of the hydrogen covered areas the tip has to move towards the surface in order to maintain a constant current. Therefore, the change in appearance is not a topographical effect but mainly caused by the modification in the electronic structure, the suppression of the Gd surface state.

The special behavior of the adsorption process may be understood by examining the areas which are labeled A, B, and C. In region A there are lattice dislocations even for the clean Gd surface which are marked by arrows (more easily to be seen in Fig. 4.13 marked by the ellipse). These are the starting points for the adsorption process. In region B it can be observed that the spreading of the adsorption is strongly suppressed at step edges due to surface steps on the tungsten substrate. The further evolution of this process is demonstrated in region C. Once the adsorption of hydrogen has started the suppression of the surface state then extends over the whole area. This behavior is also shown in Fig. 4.11b. It is the same area as in Fig. 4.11a but with an additional exposure of 0.8 L resulting in a total amount of 1.0 L hydrogen. The areas which were already affected after 0.2 L have spread out. Additionally, the adsorption occurred on another island (region D).

For comparison with results obtained by photoelectron spectroscopy one should keep in mind that the photoemission experiment averages over these regions; so

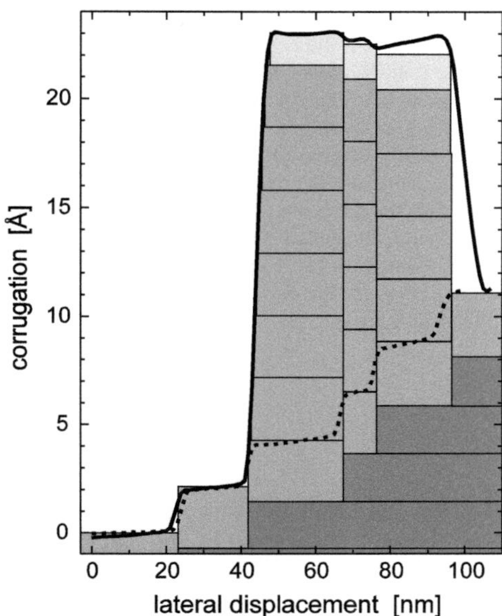

Fig. 4.12 Line section (see lines "*a*" and "*b*" in Fig. 4.11a) of a Gd island (*solid line*) and the Gd monolayer on-top of the tungsten substrate (*dotted line*). Tungsten is marked by *dark gray*, Gadolinium by *gray*, and the "electronic height" due to the surface state by light *gray shaded areas*. In order to minimize the surface energy the Gd island tries to create a flat surface resulting in an inverted step height behavior on-top of the island. Reprinted with permission from [3]. Copyright (1999) by the American Physical Society

both the peak of the Gd surface state and the hydrogen induced feature are coexistent as observed in the spectra in Fig. 4.2.

The adsorption of hydrogen seems to occur in two steps. Hydrogen is first adsorbed at surface imperfections and second, starting from these points, the adsorption spreads out to the step edges which form boundaries for the further process [3, 7, 14].

The steps of the tungsten substrate are even visible on-top of the Gd islands. This observation is caused by the different separations of the W(110) and Gd(0001) layers being 2.23 and 2.89 Å, respectively. Between the islands, the whole substrate is covered with a Gd monolayer; therefore, the step height of the monolayer is the same one as for W(110) as shown by the dotted line section in Fig. 4.12 which corresponds to line "a" in Fig. 4.11a. The solid line section (cf. line "b" in Fig. 4.11a) demonstrates the inverted step height behavior on-top of the Gd island, i.e. the tungsten terrace being one atomic layer *higher* appears *lower* on the island surface. An analogous observation was made for Gd/W(100) [15]. The surface of the island tries to be atomically flat in order to reduce its surface energy. Therefore, the height decreases from the left to the right from 7 to 5 layers. To the "topographic" height (see gray shaded area) the "electronic" height of the surface state (see light gray shaded areas) must be added. This is due to the large fade-out length into the vacuum of the d_{z^2}-like surface state accompanied by a relatively high local density of states above the surface. In the STM experiment operating in the constant-current mode the tip has therefore to be moved upwards. The tungsten substrate is marked by dark gray shaded areas.

The same area as in Fig. 4.11 is shown in Fig. 4.13. At the beginning of the scan (bottom), the total exposure so far was 1.6 L dosed during the past 1 h.

Fig. 4.13 The same area as in Fig. 4.11. At the beginning of the scan (*bottom*), the dosage was 1.6 L. During the scan, 1 L of hydrogen was additionally offered (marked by the *black arrow*). Now, unconnected areas are present (see *inset*). Reprinted with permission from [3]. Copyright (1999) by the American Physical Society

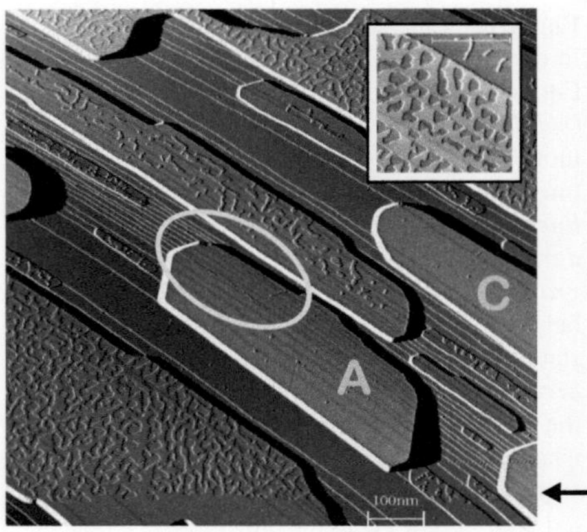

During the scan, the sample was exposed within a few seconds to an additional amount of 1 L (marked by the black arrow). Now, the islands containing regions A and C present a suppressed surface state on top of the whole island. The lattice dislocation already present in the clean Gd island is marked by an ellipse. The large island in the lower left corner shows a domain-like modification due to the additional hydrogen exposure. After the additional exposure the whole island exhibits a suppression of the surface state. The exposure time for the additional 1 L of hydrogen was substantially shorter than the previous time for 1.6 L. This observation indicates that the adsorption process changes with the dosage.

In order to determine the collision rate, one can estimate the flux, i.e. the total number of collisions per unit surface area, by $p/\sqrt{2\pi m k_B T}$ [16] with p being the pressure, m the mass of a molecule, and T the temperature. Considering the ion gauge correction factor the collision rate at room temperature for 1 L is about 30 hydrogen molecules per nm^2. On the clean surface the presence of two or more hydrogen molecules *simultaneously* seems to be necessary to cause adsorption. This conclusion is corroborated by the observation that prior to the supplementary exposure of 1 L the hydrogen covered areas are connected but afterwards the additionally created regions are mainly isolated from each other (see inset). This means that on areas of the surface being clean and of high crystallographic order the adsorption of hydrogen cannot be carried out via a spreading process but only by a simultaneous presence of at least two hydrogen atoms at one time. On the Gadolinium monolayer no alterations can be observed. Significant modifications take place on the lower islands as will be shown below.

Figure 4.14a shows the topography of approximately five monolayers (ML) Gd evaporated on the W(110) substrate held at 530 K. Since the Gd islands are atomically flat and the substrate exhibits several one-atomic high steps below the island surface, the local coverage Θ_{loc} decreases for every island from the left towards the right edge. Simultaneously with the topography, the *dI/dU* spectrum was measured at every pixel of the scan. Figures 4.14b and c show maps of the differential conductance *dI/dU* for different sample bias: (b) $U = +0.7$ V and

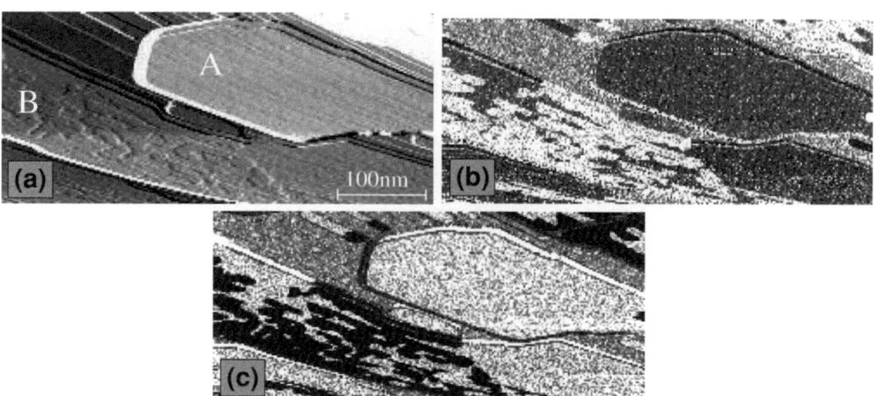

Fig. 4.14 a Topography of a Gd island system being exposed to hydrogen. The clean Gd surface is marked by *A*, the hydrogen affected island by *B*. **b, c** Maps of the differential conductance *dI/dU* for a bias voltage of $U = 0.7$ V in (**b**) and $U = -0.3$ V in (**c**) being *gray* coded. The *bright areas* in the lower part are caused by hydrogen covered areas (**b**) whereas they look dark for negative voltages (**c**). This behavior is obvious in connection with the *dI/dU* spectra in Fig. 4.10. Reprinted with permission from [3]. Copyright (1999) by the American Physical Society

(c) $U = -0.3$ V. The differential conductance is gray-coded, i.e. the higher the local *dI/dU* signal the brighter a location appears. At a sample bias $U = +0.7$ V the tunneling current is dominated by electrons which tunnel from the tip into unoccupied sample states with a binding energy of +0.7 eV. A comparison with the topographic data of Fig. 4.14a reveals that at this particular binding energy the differential conductance above the Gd monolayer is higher than above those parts of the islands which have not been affected by hydrogen. Beside few small bright pixels, the *dI/dU* signal measured above the Gd island is uniform and therefore independent of the local coverage. No contrast was found in the *dI/dU* maps on the Gd island at any sample bias (cf. Fig. 4.14b and c) in the voltage range under study (-0.6V $\leq U \leq +0.9$V). A high contrast occurs for the island B in the lower part of the image. The bias voltage of 0.7 V was chosen due to the large differential conductance for hydrogen covered areas (cf. Fig. 4.10). The bright parts in Fig. 4.14b are therefore areas which have been affected by hydrogen. The dark appearance of the island A in the upper right part points to a clean surface. The tunneling spectra reveal (see Fig. 4.10) that at a bias voltage of -0.3 V the differential conductance for hydrogen affected areas is significantly reduced. This inversion of the contrast is shown in Fig. 4.14c with black areas indicating adsorbed hydrogen.

The left part of Fig. 4.15 shows an enlarged image of a three monolayer high Gd island corresponding to the system presented in Fig. 4.13. Only those areas with hydrogen already adsorbed on exhibit small clusters. This observation is due to the onset of Gadolinium hydride being formed which will be discussed in Chap. 2. The line section (see line "a") shows a uniform height of these clusters being about 4 Å (middle). The suppression of the Gd surface state caused by

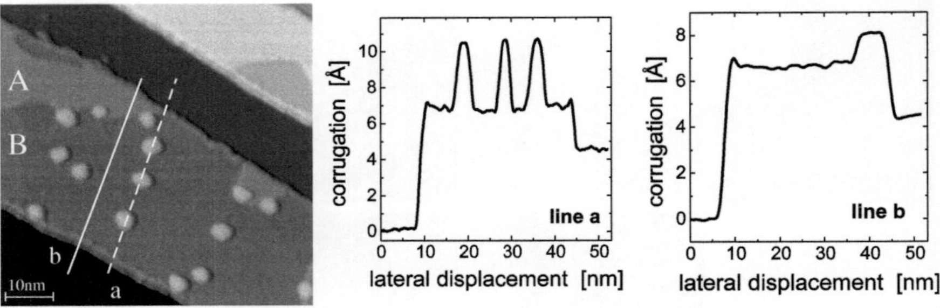

Fig. 4.15 *Left* constant-current topograph ($U = -0.3$ V, $I = 0.03$ nA) of a three monolayer high Gd island on W(110) after an hydrogen exposure of 1.6 L and subsequently of 1 L (cf. Fig. 4.13). Clean Gd is marked by A, the hydrogen affected areas by B. *Middle* line section of the clusters (line *a*). The width and the height are nearly uniform and amount to 35 and 4 Å. respectively. *Right* the suppression of the surface state due to hydrogen adsorption results in a "collapsed" looking area. The depth of this purely electronically induced depression amounts to about 1.4 Å as being obvious from the line scan (line *b*). Reprinted with permission from [3]. Copyright (1999) by the American Physical Society

hydrogen adsorption and occurring as a collapsed region (labeled by B, the clean Gd surface is marked by A) is demonstrated by the line section in the right part of Fig. 4.15 (see line "b"). The depth of this purely *electronically* induced suppression already discussed above amounts to 1.4 Å. This value is smaller than for the suppression on high Gd islands. The Gd surface state is fully developed even for at least four monolayer high areas [17]. The island with a height of three atomic layers possesses a surface state with a reduced fade-out length into the vacuum. The suppression due to hydrogen adsorption therefore seems to be smaller than for high Gd islands.

The question arises whether or not the former findings are peculiar for Gd islands, i.e. depend on the sample morphology, or are also characteristic for the element gadolinium. Therefore, hydrogen adsorption was additionally investigated for thick smooth Gd films. In Fig. 4.16a the corresponding topography is shown

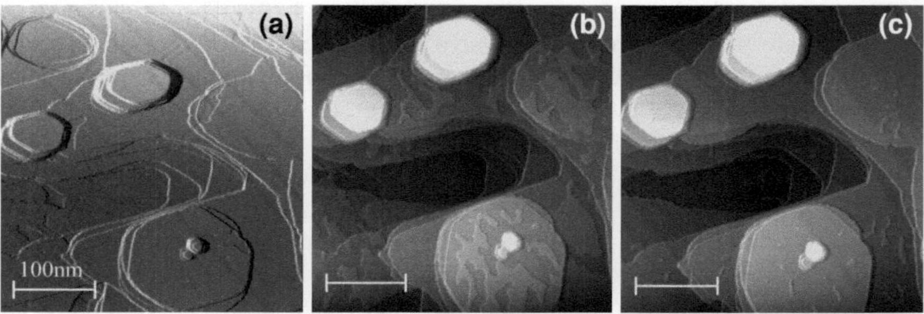

Fig. 4.16 Topography of a thick (about 40 monolayers) smooth Gd film being exposed to **a** 0.3 L, **b** 3.6 L, and **c** 9 L hydrogen. The adsorption starts at surface imperfections. Reprinted with permission from [3]. Copyright (1999) by the American Physical Society

after hydrogen exposure of 0.3 L. Again, the adsorption starts at the step edges and surface imperfections. As for Gd islands the affected areas are connected (cf. Fig. 4.11a). The hydrogen "island-like" formation was also deduced from photo-emission experiments for H/Gd(0001) [5] and H/Be(0001) [18] by determination of peak heights from hydrogen induced states as a function of dosage. Further exposure (3.6 L in Fig. 4.16b) demonstrates the same behavior in the adsorption process, a lagoon-like appearance. After a dosage of 9 L (see Fig. 4.16c) nearly the whole Gd film is covered with hydrogen. These observations point to no significant differences in the adsorption process of hydrogen on islands and on smooth films.

The influence of the bias voltage on the topography is demonstrated in Fig. 4.17. At negative voltages ($U = -0.3$ V in Fig. 4.17a) the hydrogen affected areas exhibit a reduced apparent height due to the drastically reduced LDOS

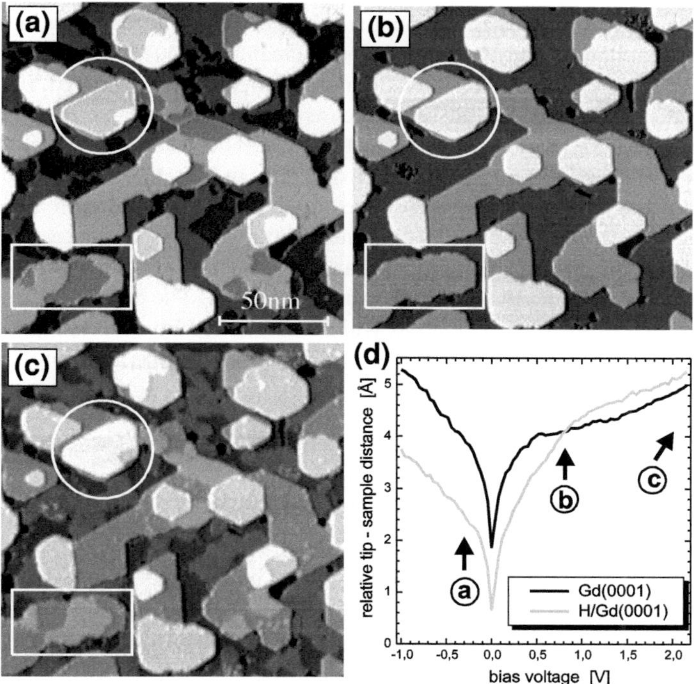

Fig. 4.17 Topography of a thick Gd film being exposed to about 1 L of hydrogen as a function of the bias voltage. For negative values ($U = -0.3$ V in **a**) the hydrogen affected areas appear deeper due to their very small differential conductance whereas an inversion can be observed ($U = 3.2$ V in **c**) for high positive voltages. The contrast vanishes at low positive voltages ($U = 0.8$ V in **b**). This behavior is obvious for the island being nearly completely covered with hydrogen (marked by the *circle*) and the region with a partial coverage (marked by the *rectangle*). The *scale bar* corresponds to 50 nm. **d** Dependence of the relative tip–sample distance on the bias voltage for uncovered Gd(0001) and hydrogen affected areas ($I = 0.2$ nA). The voltages which the images are taken at are marked by *a*, *b*, and *c*. It is obvious that at 0.8 V no difference between these two areas are present. Reprinted with permission from [3]. Copyright (1999) by the American Physical Society

(cf. Fig. 4.10). Using moderate positive bias voltages ($U = 0.8$ V in Fig. 4.17b) it is possible to obtain equal apparent heights for the clean Gd(0001) surface and hydrogen affected areas. This observation is caused by the drastic increase of the LDOS for hydrogen on Gd(0001) towards higher voltages as already demonstrated in Fig. 4.10. For high positive voltages ($U = 3.2$ V in Fig. 4.17c) an inversion of the corrugation, i.e. the hydrogen affected areas now appearing higher than for negative voltages, is present as can additionally be deduced from the shape of the LDOS curves. Therefore, the depth of the so-called "collapse" (cf. Fig. 4.15) is not uniform and depends significantly on the bias voltage. For a direct determination of this dependence $z(U)$ measurements were carried out by scanning the sample bias at a constant tunneling current with the feedback left on and measuring the z-feedback signal versus sample voltage. With the z-calibration, obtained at monoatomic Gd(0001) steps, one gets the relative tip displacement as function of sample bias. The determination of the displacement was obtained in a current imaging mode at a region of the sample that consists of both clean Gd(0001) as well as hydrogen covered areas. For each curve, about 50 measurements were averaged. Figure 4.17d shows the relative tip–sample distance as a function of the bias voltage for uncovered Gd(0001) and hydrogen affected areas. Those voltages which the images are taken at are marked and correspond to the ones in Fig. 4.17a–c. It is obvious that at negative and low positive bias voltages up to 0.8 V the hydrogen affected areas exhibit a significantly reduced apparent height whereas at increasing positive voltages they appear a little bit higher than the clean Gd(0001) regions. At 0.9 V no differences between the two types of areas are present. The dependence on the bias voltage is clearly visible in the areas which are marked by a circle representing an island which is nearly completely covered with hydrogen and a rectangle showing a partially affected island, respectively.

4.2 Hydrogen Incorporation in Gadolinium

In the following Chapter the discussion of the behavior of thin Gd films is continued being exposed to hydrogen but for significantly higher doses [19] which results in a penetration of the hydrogen atoms into the bulk material.

Hydrogen incorporation in thin films can produce an extremely high out-of-plane expansion due to the clamping of the thin film to the substrate [20]. Within the solid solution phase the expansion can be predicted by linear elastic theory [21]. However, deviations from linear elastic theory were reported for high H concentrations due to the onset of plastic deformation in the film [22, 23].

Two different processes have been proposed for plastic deformation and stress relaxation in thin films [22, 23]. The emission of extrinsic dislocation loops during hydride precipitation as observed experimentally in bulk material [24] and the glide of dislocation segments that originate from misfit dislocations in the vicinity of the interface between the film and the substrate [25]. The strain energy of the film increases during hydrogen incorporation since the misfit between the adhering film and the substrate increases with H concentration. Thus, above a certain H concentration the formation of a misfit dislocation is energetically favored.

As already discussed in Sect. 3.1 (Chap. 3) the electronic structure of clean and well-ordered Gd(0001) exhibits a pronounced surface state [7]. After hydrogen adsorption the surface state disappears, and a significant decrease in the differential conductance occurs at negative bias voltages. This means that the hydrogen covered areas of the sample appear lower in the STM images obtained in the constant-current mode (cf. Sect. 4.1.2). Figure 4.18 shows a Gd(0001) surface after hydrogen exposures of 5 L (a), 10 L (b), 20 L (c,d). After 20 L exposure (c) only a small area remains free of hydrogen and since this area still has the Gd surface state it appears slightly higher. In the following STM image (see Fig. 4.18d), taken 40 min later, the elevated area has been increased. The suppression and reappearance of the Gd surface state on the *same* terrace is indicated by white circles in Fig. 4.18b–d. Its existence can be deduced from the depression followed by appearing of the apparently higher areas. The reappearance of the Gd surface state shows that hydrogen has disappeared from the surface. Hydrogen has a large negative heat of solution in rare earth metals, and therefore it dissolves in interstitial sites even at very low gas pressures [26]. Thus, it is concluded that the hydrogen being initially adsorbed *on* the surface has diffused *into* the interior of the film.

After an exposure of 20 L the surface topography changes in certain areas. Figure 4.18e shows such a location of the sample; the circle marks the same area as the circle in Fig. 4.18d. Two different features were observed: first, a large number of disc-like islands; second, occasionally ramp-shaped features. The disc-like islands have a diameter of 35 Å and a height of about 3 Å (see Fig. 4.19a). The patterned area appears to be rough compared to the smooth Gd or H/Gd surface on each terrace and is induced by the surface modification which will be

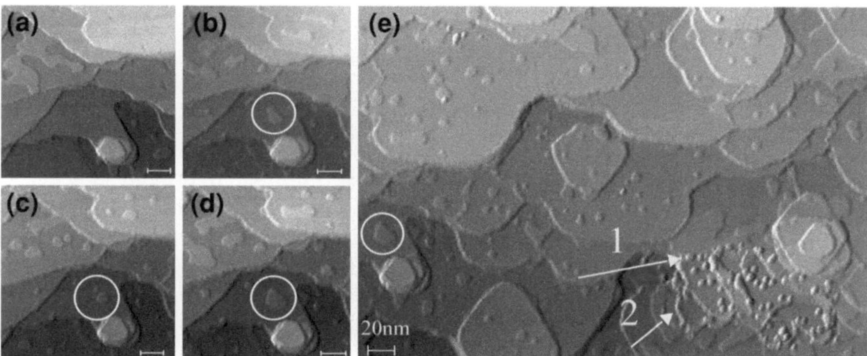

Fig. 4.18 STM images of a Gd(0001) film after total hydrogen exposures of 5 L (**a**), 10 L (**b**), and 20 L (**c, d**); all *bars* represent 20 nm. The Gd surface state is detectable as areas appearing higher on the *same* terrace in (**a**) whereas in (**c**) it is almost disappeared. It reappears after 40 min, as can be seen in (**d**) indicating the removal of hydrogen from the surface. After 20 L, two different types of surface pattern appear. **e** Small islands with disc-like shape *1* and ramps *2*. Reprinted with permission from [19]. Copyright (2000) by the American Physical Society

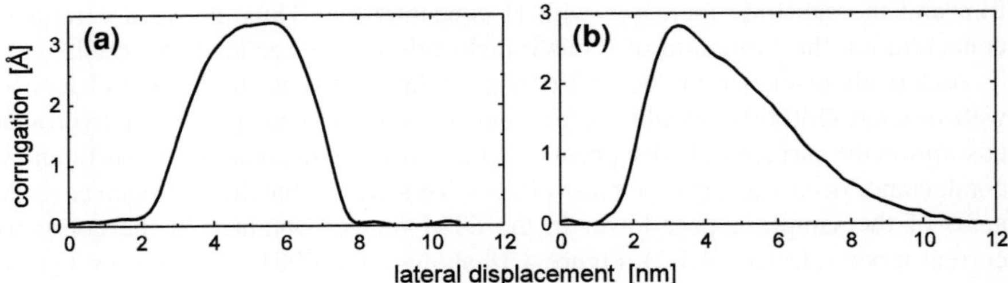

Fig. 4.19 Height profiles of small disc-like islands (**a**) showing steps at their sides, and a ramp (**b**), taken perpendicularly to the straight line. Here a steep step is followed by an inclination over several nanometers to the base level. Note the different scales for lateral displacement and corrugation. Reprinted with permission from [19]. Copyright (2000) by the American Physical Society

discussed below. The height profiles show that the edges of the island have steep steps. A height profile taken along the sloped region of a ramp is shown in Fig. 4.19b. The line scan is taken from the left. The steep step occurs at the left side. The inclination extending from the step to the right, covers several nanometers until it reaches the base level. Both, islands and ramps are elevated [by about the distance of the (0001) planes] compared to the level of the Gd terrace and are therefore generated by *local material transport to the surface.*

The ramps only appear in regions where disc-like island formation has already taken place and seem not to originate from chains of close packed disc-like islands since such an elevation should result in a step at all border lines, as shown in Fig. 4.19a. In contrast a level inclination was found on one side and a step on the other side. Thus, the two surface pattern have a different origin.

To conclude, ramps and island formation verify that plastic deformation of the film occurs during hydrogen incorporation.

4.3 Coadsorption Characteristics

In order to get a deeper insight into the coadsorption of hydrogen and carbon monoxide (CO) on gadolinium it is necessary to understand what happens when exposing the Gd surface to each single type of molecules. The behavior of hydrogen was already discussed in Sect. 4.1; the adsorption of CO will first be described in the following.

4.3.1 Gadolinium Exposed to Carbon Monoxide

The adsorption process of CO on gadolinium is strongly depending on the substrate temperature. Exposing Gd to CO results in a behavior which is illustrated in Fig. 4.20. The left image is related to the adsorption of 0.8 L CO at low

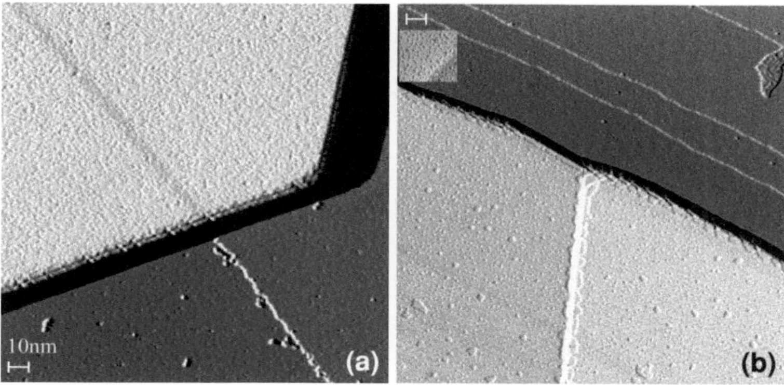

Fig. 4.20 Topography of Gd islands being exposed to CO at 90 K (**a**) and 293 K (**b**). The surfaces of the Gd islands are differently structured due to a molecular adsorption at low temperature and to a dissociation of the CO molecules at room temperature. Both *scale bars* correspond to 10 nm. Reprinted with permission from [3]. Copyright (1999) by the American Physical Society

temperature (90 K), the right one corresponds to 0.6 L CO at room temperature. The surfaces of the Gd islands are differently structured. The inset in Fig. 4.20b shows, on the *same* scale, the Gd surface after an oxygen dosage of 0.08 L. Whereas at low temperature the adsorption seems to occur molecularly being in good agreement, e.g., with investigations on CO adsorption on Fe(110) [27] and on Co(0001) [28], dissociation on this Gd surface takes place at 300 K as already shown by Searle et al. [29] and additionally, e.g., for CO on Fe(100) surfaces [30, 31]. The similarity between CO and O_2 adsorption is a strong hint to the dissociation of the carbon monoxide molecule at room temperature. Direct evidence will be given by means of photoelectron spectroscopy (see below).

An exposure sequence obtained at room temperature is presented in Fig. 4.21. For small dosages (1 L in Fig. 4.21a) the surface is relatively smooth with single localized areas being depressed. For higher exposures (20 L in Fig. 4.21b) the appearance has been changed. As will be discussed in more detail below the rough surface points to the creation of carbonate species.

4.3.2 Coadsorption of Hydrogen and Carbon Monoxide

For a better understanding of catalytic processes the coadsorption of hydrogen and carbon monoxide was investigated. For this purpose the Gd surface was exposed to small amounts of hydrogen with a subsequent dosage of CO [3, 32]. The basic steps of this process were determined with a sample exposed to 1 L hydrogen being adsorbed on a Gd(0001) surface. This system was subsequently exposed to CO with dosages up to 0.81 L. The photoemission spectra are presented in Fig. 4.22. With increasing CO exposure it can be observed that the Gd surface

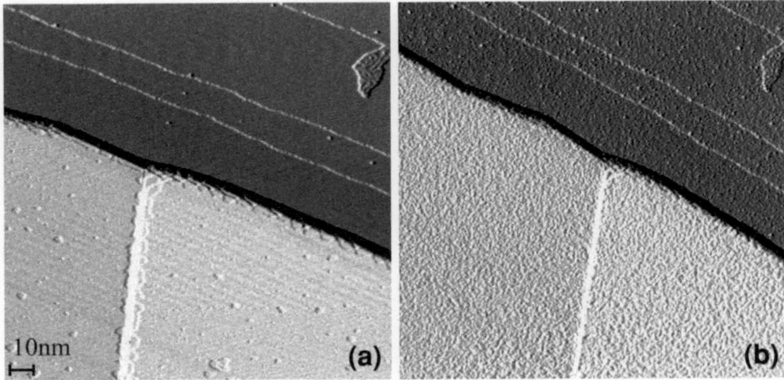

Fig. 4.21 For small exposures at room temperature (1 L in **a**) the surface looks relatively smooth whereas for higher dosages (20 L in **b**) it becomes rough. Reprinted with permission from [3]. Copyright (1999) by the American Physical Society

Fig. 4.22 Photoemission spectra taken in normal emission with $hv = 16.85$ eV for a smooth Gd film being pre-exposed to 1 L hydrogen as a function of CO dosage. With increasing offer the hydrogen induced peak at a binding energy of 4 eV completely disappears whereas around 2 eV a broad structure shows increasing intensity. Reprinted with permission from [3]. Copyright (1999) by the American Physical Society

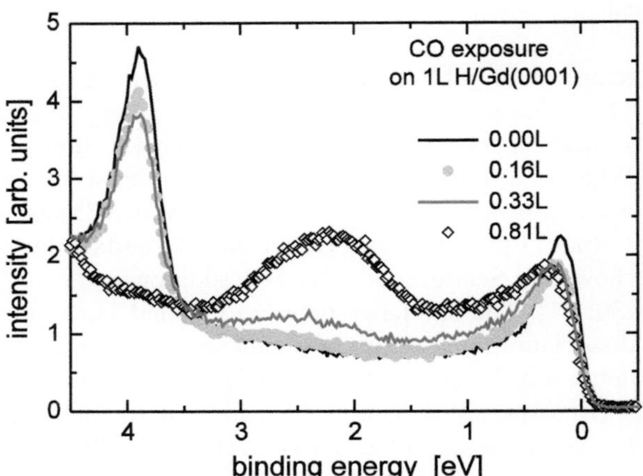

state near the Fermi level remains nearly unaffected apart from a small decrease in intensity when initiating the CO exposure and a slight shift to higher binding energies. Additionally, the hydrogen induced feature at 4.0 eV binding energy looses intensity and is absent after 0.81 L CO. In the energy region around 2 eV an additional structure appears with increasing CO exposure; it is not apparent at low coverages.

It was shown [33] for uranium (U), possessing a related electronic structure ([Rn] $5f^3$ $6d^1$ $7s^2$) to gadolinium ([Xe] $4f^7$ $5d^1$ $6s^2$), that after CO adsorption two features appear; one structure at 2.4 eV binding energy is due to C $2p$ emission from uranium carbide (UC) and the other one at 2.2 eV from U caused by the existence of oxygen.

Due to the similar electronic structure to uranium, the peak intensities for Gd as substrate were analogously determined at 2, 2.4, and 4 eV as a function of CO

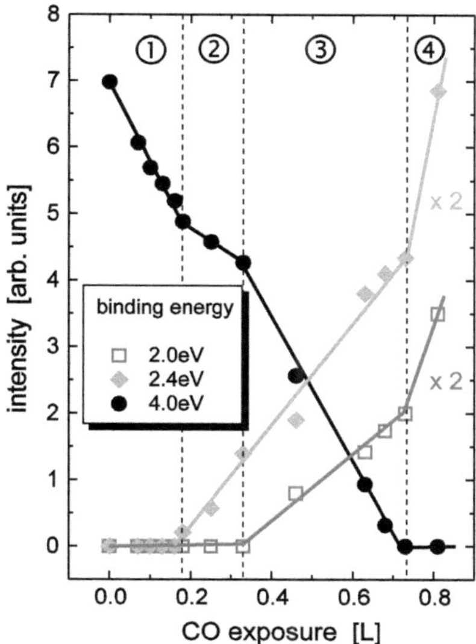

Fig. 4.23 Peak intensities as a function of CO dosage. The *shapes* of the curves point to four regions with different behavior in the adsorption process (the details are discussed in the text). Reprinted with permission from [3]. Copyright (1999) by the American Physical Society

exposure (see Fig. 4.23). The procedure consists of the subtraction of a linear background and a fit with Gaussian functions. It is obvious that the diagram can be classified by four regions, i.e. the adsorption process up to 0.81 L consists of four steps.

At the beginning the hydrogen induced intensity decreases very fast whereas the dissociatively adsorbing CO causes no features in the spectra neither for the C $2p$ state at 2.4 eV nor for the oxygen caused Gd state at 2.0 eV. In the next step the loss of intensity of the hydrogen state is diminished, the emission of the carbon state starts, and no feature caused by oxygen is present. Step 3 consists of the decrease in hydrogen induced intensity with the same slope as in step 1, increasing carbon induced intensity with the same slope as in step 2, and the onset of oxygen induced intensity from a Gd state. In step 4 no hydrogen induced intensity is present and the carbon induced and oxygen caused states show a strengthened increase in intensity.

In the following the chemical reactions will be described being induced by increasing CO exposure; the superscript *gas* means gaseous, *ads* means adsorbed in any way. For a better comparison all formulae were referenced to a total of four CO molecules.

During the first step only hydrogen is removed, no additional species are adsorbed; all products are therefore gaseous pointing to the creation of formaldehyde which can directly be created via a gas phase reaction:

$$\textbf{Step 1}: \qquad 8\,H^{ads} + 4\,CO^{gas} \longrightarrow 4\,H_2CO^{gas}$$

The ratio of eight H atoms to four CO molecules for the complete removal of hydrogen can *independently* be estimated by the following consideration.

It was shown by Li et al. [5] that 2 L H_2 are necessary in order to get a full coverage of Gd(0001) with an (1×1) overlayer. Due to the dissociative adsorption of H_2 one can estimate, using the Gd lattice constants, that the surface is covered with about eight H atoms per nm^2. In this investigation Gd was exposed to 1 L H_2, i.e. the density amounts to four H atoms per nm^2. The extrapolation in region 1 (see Fig. 4.23) results in 0.6 L CO being necessary to remove all hydrogen. 1 L CO is equivalent to a collision rate of 3.5 molecules/nm^2 (cf. the estimation for hydrogen in Sect. 4.1), therefore 0.6 L CO give a rate of about two CO molecules/nm^2. This means that two CO molecules in relation to four H atoms cause a complete removal of hydrogen as stated above (see reaction scheme of step 1).

In the next step the reaction changes; only half a rate of hydrogen is removed as can be seen by the change in slope for the H induced feature in Fig. 4.23. Thus, another reaction seems to occur. The additional amount of CO causes the creation of carbon on or near the surface but no oxygen atom is adsorbed. The reaction can therefore be written as

$$\textbf{Step 2:} \qquad Gd + 4\,H^{ads} + 4\,CO^{gas} \longrightarrow 2\,CO_2^{gas} + CH_4^{gas} + GdC^{ads}$$

The formation of Gadolinium carbide was also observed for U exposed to CO [33].

During further exposure hydrogen is removed with the same rate compared to the initial step 1. The rate of carbon creation remains constant, but now oxygen is adsorbed creating a subsurface oxide and causing emission from Gd^{2+} states as already discussed for oxygen adsorbed on Gd [34]. This reaction can be described by

$$\textbf{Step 3:} \qquad 3\,Gd + 8\,H^{ads} + 4\,CO^{gas} \longrightarrow CO_2^{gas} + 2\,CH_4^{gas} + GdC^{ads} + 2\,GdO^{ads}$$

At the end of step 3 the *whole* amount of hydrogen has been removed from the surface.

The next step is characterized by the creation of carbon and Gd suboxide through direct CO dissociation:

$$\textbf{Step 4:} \qquad 4\,Gd + 4\,CO^{gas} \longrightarrow 4\,GdC^{ads} + 4\,GdO^{ads}$$

As every CO molecule is transformed now the increase in the peak intensities around 2 eV binding energy becomes even more pronounced.

In order to determine the influence of higher CO dosages a Gd film covered with the equivalent of 0.4 L hydrogen was subsequently exposed to CO by up to 3.5 L. These spectra are presented in Fig. 4.24. In the low coverage regime the results are comparable with the system discussed above. In contrast, the spectrum for 3.5 L CO appears significantly different. No feature at the Fermi level is present anymore and the one around 2 eV binding energy shows a double peak structure. The peak intensities as a function of CO exposure are given in Fig. 4.25. It is obvious that the same mechanism takes place. The corresponding STM images taken for 0 and 0.6 L CO, shown with an expanded scale in Fig. 4.28, demonstrate the removal of hydrogen (see discussion below).

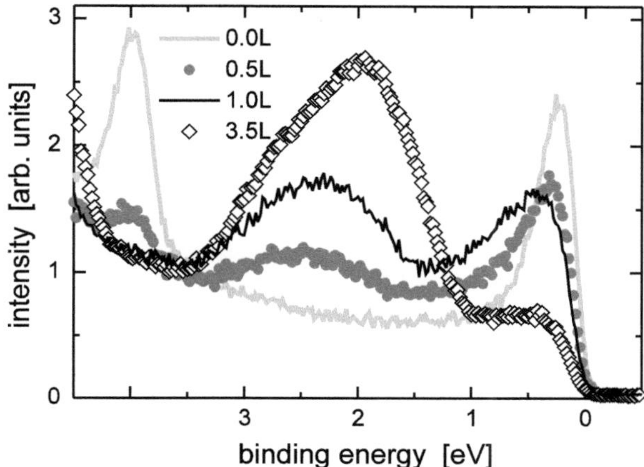

Fig. 4.24 Photoemission spectra for 0.4 L H/Gd(0001)/W(110) with increasing CO dosage (cf. Fig. 4.22)

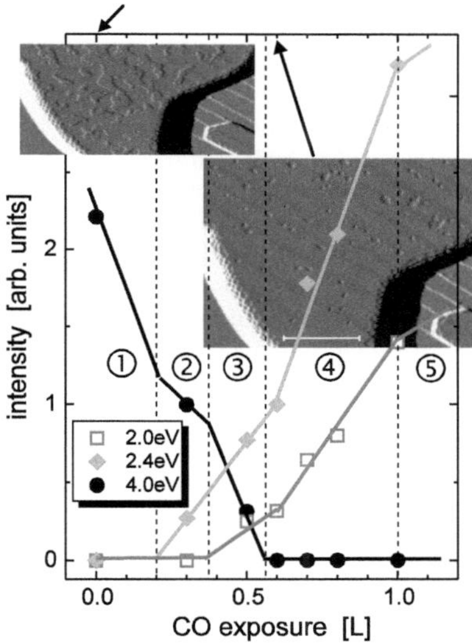

Fig. 4.25 Intensities of the hydrogen induced structure at 4 eV, the oxygen caused Gd state at 2.0 eV, and the C $2p$ state at 2.4 eV as a function of CO dosage. The adsorption process can be divided into five steps. The corresponding topography is shown for 0 L CO and 0.6 L CO as *insets*. The removal of hydrogen can be observed due to the reappearance of the Gd surface state. Reprinted with permission from [3]. Copyright (1999) by the American Physical Society

Above 1 L an additional regime 5 occurs. This is illustrated by an expanded scale in Fig. 4.26; the corresponding STM image is given for a CO dosage of 3.5 L. The increasing intensity becomes less pronounced compared to region 4. The change in the electronic properties is additionally visible in the energy range of O $2p$ states around 6 eV binding energy. Figure 4.27 shows the spectra taken after CO exposures of 0.8 and 2.8 L, respectively. Whereas for the low dosage a single broad peak can be observed pointing to oxygen in Gd subsurface oxide [35], at higher exposures a double-peak structure appears which is characteristic for Gd_2O_3 [34]. Analogous experiments were carried out with CO_2 on Y(0001) showing that the dissociation of CO_2 results in similar features [36–38]

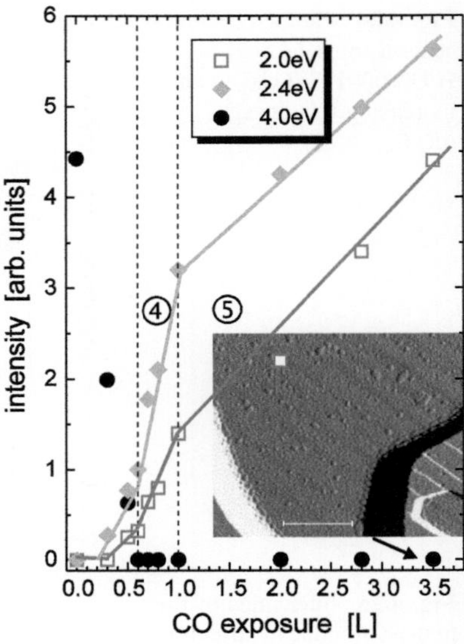

Fig. 4.26 Same as in Fig. 4.25 but for higher coverages. The corresponding STM image is taken for a dosage of 3.5 L CO. Reprinted with permission from [3]. Copyright (1999) by the American Physical Society

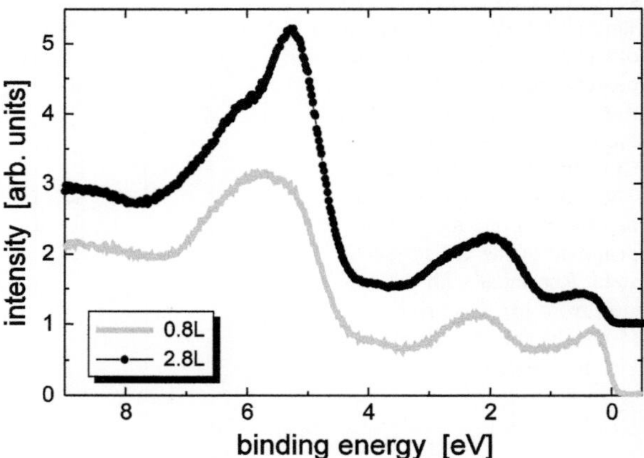

Fig. 4.27 Gd films being pre-exposed to hydrogen for different CO dosage. The broad peak around 6 eV binding energy is due to oxygen. The double-peaked structure for higher dosages points to the formation of Gd_2O_3. This is in agreement with the drastically decreased intensity near the Fermi level

It is known [39] that the sesquioxide of rare earth metals RE_2O_3 are able to induce the transformation of CO to CO_2 which then form a stable carbonate-like species at the surface. In Gd_2O_3 Gd is present in the oxidation stage Gd^{3+} thus having lost all valence electrons; the intensity near the Fermi level is drastically decreased (see Fig. 4.27). The significantly different structure around 2 eV binding energy is attributed to emission from the carbonate-like species as already supposed by Searle et al. [29]. Taking these considerations into account the further step in the process can be described by

Step 5 : $6\,GdO + 4\,CO^{gas} \longrightarrow 2\,GdC^{ads} + 2\,(Gd_2O_3 \cdot CO_2)^{ads}$

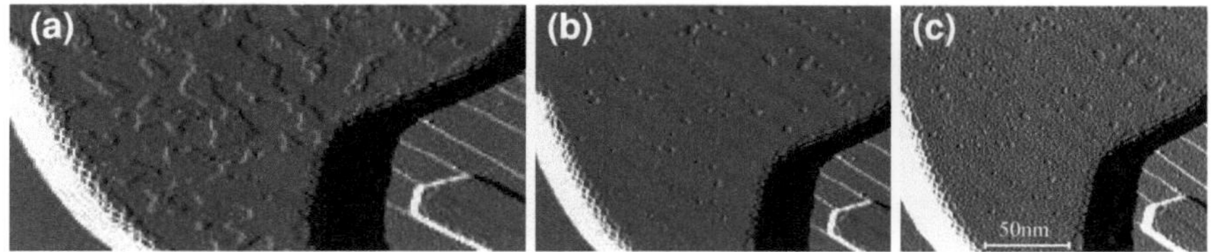

Fig. 4.28 Enlarged images of the STM pictures in Figs. 4.25 and 4.26 taken at CO dosages of 0 L (**a**), 0.6 L (**b**), and 3.5 L (**c**). Reprinted with permission from [3]. Copyright (1999) by the American Physical Society

It should be noted that a part of the carbonate-like species may desorb as CO_2^{gas}.

The "visualization" of this process was carried out via scanning tunneling microscopy. For this purpose Gd islands with hcp(0001) surfaces were exposed to 0.4 L hydrogen. The affected areas can be seen in Fig. 4.28a (cf. additionally Fig. 4.25). CO was subsequently dosed to this system. After 0.6 L CO exposure no collapsed areas are observable (see Fig. 4.26) pointing to no hydrogen remaining on the surface. At this point region 4 is reached, and the complete hydrogen adsorbate is removed by CO. The single areas with a reduced apparent height are caused by contamination during Gd film preparation. They are also observable in Fig. 4.28a. Figure 4.28c shows the sample which was exposed to 3.5 L CO. In step 5 carbonate species are created which cause the surface roughness.

This direct comparison, i.e. the *visualization* of the changes in the electronic properties caused by chemical reactions at the surface, demonstrates the advantage of the combination of photoelectron spectroscopy with scanning tunneling microscopy and spectroscopy.

4.4 Oxygen on Rare Earth Metal Surfaces

First it will be reported on the investigation of oxygen adsorbed on rare earth metal surfaces using tunable VUV photoelectron spectroscopy [40] followed by the description of scanning tunneling microscopy experiments [3] which were used to determine the very first steps of the adsorption process.

The photoelectron spectra of dissociatively adsorbed oxygen on the different rare earth surfaces show the well-known feature at a binding energy of about 6 eV. This structure is caused by emission from the O $2p$ states. As an example, the photoelectron spectrum of 1 L oxygen on Gd(0001)/W(110) is shown in Fig. 4.29. The asymmetrical shape in the oxygen induced feature seems to indicate a splitting into $2p_{xy}$ and $2p_z$ states. The adsorbate (O $2p$) induced structure reveals a large difference in the photoelectron intensities for excitation with right (I^{RCP}) and left (I^{LCP}) circularly polarized radiation. The intensity difference means that a Circular Dichroism in the Angular Distribution of photoelectrons (CDAD) effect is present.

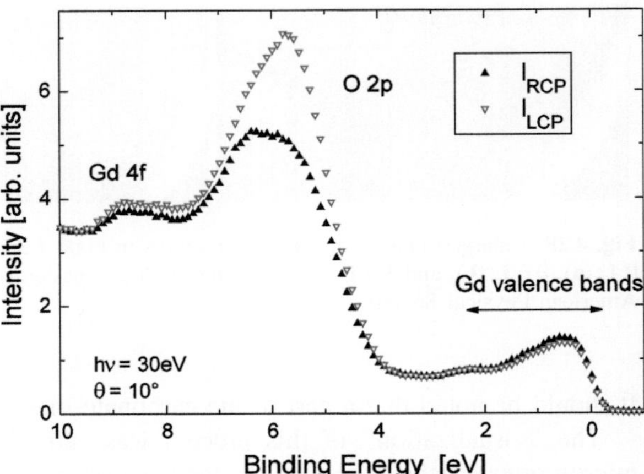

Fig. 4.29 Pair of photoelectron spectra of 1 L oxygen adsorbed on Gd(0001)/W(110) at an emission angle of $\theta = 10°$ and $hv = 30$ eV. The different intensities after excitation with right (I^{RCP}) and left circularly polarized light (I^{LCP}) reflect a CDAD effect in the O $2p$ state as well as the Gd $4f$ states. Reprinted from [40], Copyright (1996), with permission from Elsevier

The underlying background of secondary electrons exhibits no CDAD. These electrons have been inelastically scattered several times and lost their prior angular distribution. Not only the oxygen feature but also the rare earth metal valence bands show different photoelectron intensities indicating that CDAD effects do also arise for thin films and bulk material, respectively.

In order to extract quantitative information from the photoelectron spectra, an asymmetry function is defined as follows

$$A_{\mathrm{CDAD}}(\theta) = \frac{I_{\mathrm{CDAD}}}{I_0} = \frac{I^{\mathrm{RCP}}(\theta) - I^{\mathrm{LCP}}(\theta)}{I^{\mathrm{RCP}}(\theta) + I^{\mathrm{LCP}}(\theta)} \qquad (4.11)$$

where $I^{\mathrm{RCP}}(\theta)(I^{\mathrm{LCP}}(\theta))$ denotes the photoelectron intensity above the secondary electron background at the emission angle θ. The intensities are obtained by determining the peak heights above the background. There are no significant differences in the ratios determined using the peak heights and the peak areas.

Figure 4.30 shows the CDAD asymmetry of the $2p$ state of oxygen adsorbed on Nd(0001)/W(110) (top), on Tb(0001)/W(110) (middle), and on Gd(0001)/W(110) (bottom) as a function of the emission angle. The exposure was 1 L and the excitation energy $hv = 30$ eV. In all cases, the theoretically predicted [41] (see below) antisymmetrical behavior of the asymmetry $[A_{\mathrm{CDAD}}(\theta) = -A_{\mathrm{CDAD}}(-\theta)]$ is clearly visible. For higher emission angles ($|\theta| > 20°$), a nearly constant asymmetry is detected. For Nd and Tb it decreases when varying the emission angle to lower values.

The system O/Gd/W(110) exhibits a significantly different behavior for angles smaller than 20°. A pronounced feature overlaps the general shape of the asymmetry curve: an additional change of sign and a high maximum of 15% at $\theta = -10°$ (deep minimum of -18% at 10°, respectively) occurs. This observation directly points to different chemical or electronic properties in the adsorbate overlayer.

For an aligned p valence level of a free atom, the general shape of I_{CDAD} should follow $\sin 2\theta$ as shown for comparison in Fig. 4.30 [41]. The asymmetry contains

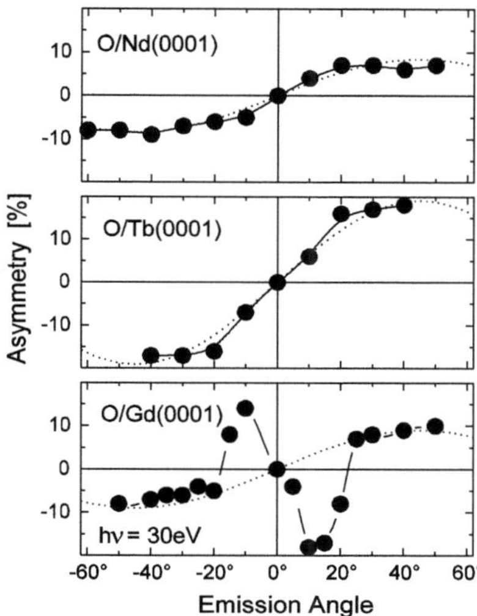

Fig. 4.30 CDAD asymmetry for oxygen adsorbed on Nd (*top*), Tb (*middle*), and Gd (*bottom*) after an exposure of 1 L ($hv = 30$ eV). For small emission angles, the asymmetry for O/Gd(0001) is very different as compared to O/Nd(0001) and O/Tb(0001). The *dotted line* corresponds to $\sin 2\theta$, the angular pattern expected for an aligned p state in isotropic space. The *full line* is to guide the eye. The error amounts to about 5%. Reprinted from [40], Copyright (1996), with permission from Elsevier

an additional angular dependence due to the normalization to the "unpolarized" intensity $I_0 = 1 + f(\theta)$ and can therefore be described by

$$A_{CDAD} = \frac{I_{CDAD}}{I_0} \propto \frac{\sin 2\theta}{1 + f(\theta)} \qquad (4.12)$$

The CDAD asymmetries determined for O/Nd and O/Tb follow this predicted behavior quite well. Refraction of the electrons at the surface vacuum barrier and a coherent superposition of directly emitted and reflected electron beams give rise to additional changes in the angular dependence of the CDAD [42]. At the low kinetic energies (<30 eV), used here, these effects cannot explain the shape of the asymmetry observed for O/Gd. The sharp increase in the CDAD at low emission angles can be explained taking into account a superposition of the $l = 1$ state with spherical harmonics of higher angular momenta to form the initial state. This directly points towards a different chemical environment in the case of O/Gd compared to O/Nd and O/Tb.

The asymmetry of oxygen on Gd(0001) was additionally determined at another photon energy of $hv = 36$ eV and a higher exposure of 15 L. The corresponding asymmetry curves are presented in Fig. 4.31. The pronounced feature at about $10°$ possesses a higher asymmetry value of -21%, compared to about -18% at $hv = 30$ eV. Additionally, a decrease in A_{CDAD} with increasing emission angle beyond $30°$ occurs. This behavior of a decrease was not found for O/Tb and O/Nd showing again the different electronic properties. After the high exposure, the general shape of the curve remains unchanged, only the values are smaller pointing to a scaling of A_{CDAD} with exposure. This observation is in agreement with the work of Weller and Sarma [43] who found a passive oxide film. This means exposing epitaxial Gd layers to oxygen results in dissociative chemisorption with initial exposure and is followed by formation of Gd oxide [34, 35, 43].

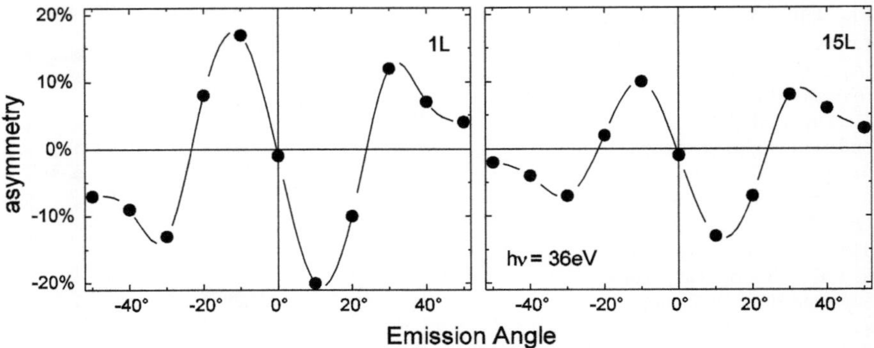

Fig. 4.31 CDAD asymmetry for oxygen on Gd/W(110) for an exposure of 1 L (*left*) and 15 L (*right*); the excitation energy was $hv = 36$ eV. For higher coverages, a reduction of the asymmetry occurs whereas the shape of the curve remains unchanged. The *full line* is to guide the eye. The error amounts to about 5%. Reprinted from [40], Copyright (1996), with permission from Elsevier

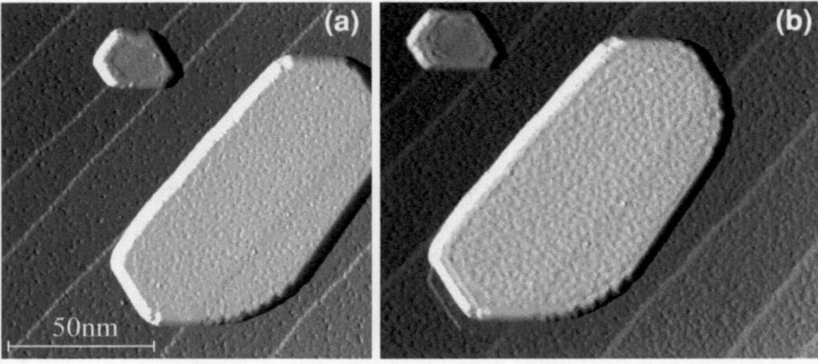

Fig. 4.32 Topography of a Gd island with the monolayer covering the tungsten substrate between the islands being exposed to 0.08 L (**a**) and 0.2 L (**b**) oxygen. Reprinted with permission from [3]. Copyright (1999) by the American Physical Society

The very first step of this adsorption process will now be discussed in more detail. Oxygen on Gd islands with the remaining monolayer in-between is shown in Fig. 4.32. In the left image the exposure amounts to 0.08 L, in the right one, showing the same area of the sample, it was increased to 0.2 L. It is obvious that in contrast to hydrogen the adsorption can be understood as a process leaving oxygen regularly distributed over the surface.

In order to demonstrate the differences between oxygen and hydrogen adsorbed on gadolinium the surface was covered with small amounts of hydrogen and subsequently exposed to 0.1 L oxygen. The topography is shown in Fig. 4.33a. The large depressed area C on-top of the island is caused by hydrogen, the small areas by oxygen. A map of the differential conductance dI/dU at -0.2 V is presented in Fig. 4.33b showing the reduced local density of states for adsorbed

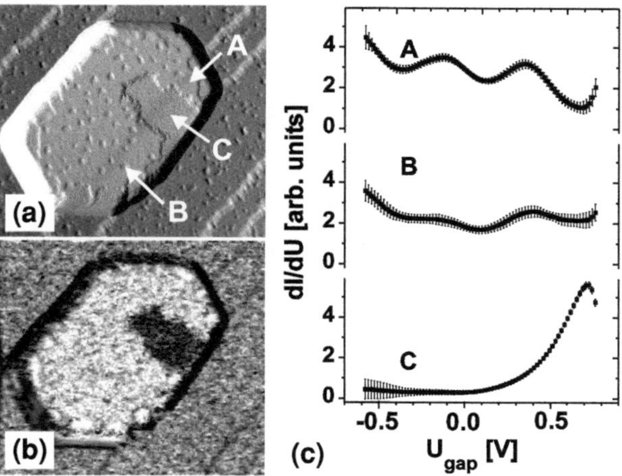

Fig. 4.33 Gd being exposed to small amount of hydrogen and subsequently to 0.1 L oxygen. **a** Topography (70 nm × 60 nm, $U = -0.7$ V, $I = 1$ nA). **b** Map of the differential conductance at -0.2 V. **c** Tunneling spectra of the Gd island for uncovered gadolinium (*A*), oxygen induced small areas (*B*), hydrogen affected area (*C*). These regions are marked in (**a**). Reprinted with permission from [3]. Copyright (1999) by the American Physical Society

hydrogen at negative bias voltages (cf. Fig. 4.14c). Tunneling spectra (see Fig. 4.33c) were recorded on-top of the island for the unaffected Gd surface (A), the oxygen induced small areas (B) as well as the hydrogen affected areas (C). Spectrum A demonstrates that on the non-covered areas the exchange split Gd surface state is present. The differential conductance within the hydrogen covered area is comparable to the one in Fig. 4.10. Curve B shows that the Gd surface state is still present at the oxygen induced regions. As already discussed in detail (see Sect. 4.3), oxygen is adsorbed *below* the Gd surface. This behavior enables that the surface state remains present but is significantly depressed. The further steps in the adsorption process have already been discussed above in connection with coadsorption of CO and H_2.

In the following the attention will be turned to oxygen adsorption on the Gd monolayer located between the Gd islands. Figure 4.34a shows a detailed image of the monolayer after an exposure of 0.08 L (cf. Fig. 4.20b). Due to their strong electronegativity the dissociatively adsorbed oxygen molecules are visible as black holes, i.e. adsorbed oxygen atoms are imaged as a depression; this behavior was already found, for example, for O/W(110) [44] and was also theoretically predicted [45]. The oxygen overlayer forms an ordered structure. This is not present (see Fig. 4.34b) after the additional amount of 0.12 L of oxygen on the surface (cf. Fig. 4.32b).

In order to determine the geometrical arrangement a more detailed image is shown in the left part of Fig. 4.35. The adsorption sites of the oxygen atoms form a hexagonal structure which is schematically outlined in the left part of Fig. 4.35. The nearest neighbor distance amounts to 13 ± 1 Å.

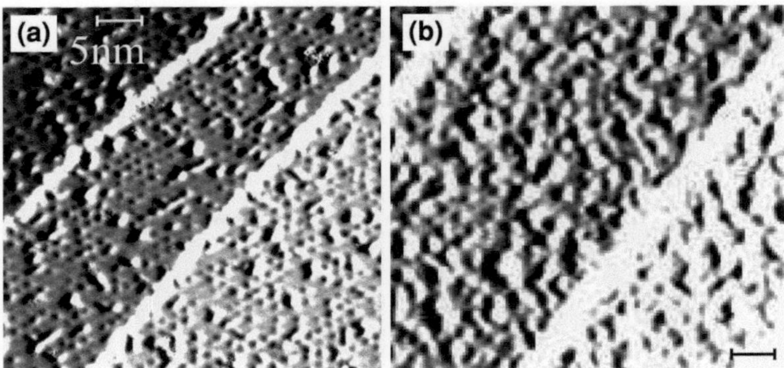

Fig. 4.34 Topography of the Gd monolayer covering the tungsten substrate between the islands being exposed to 0.08 L oxygen (**a**) and 0.2 L (**b**); the *scale bars* correspond to 5 nm. The oxygen overlayer forms an ordered structure which is not visible for higher exposures. Reprinted with permission from [3]. Copyright (1999) by the American Physical Society

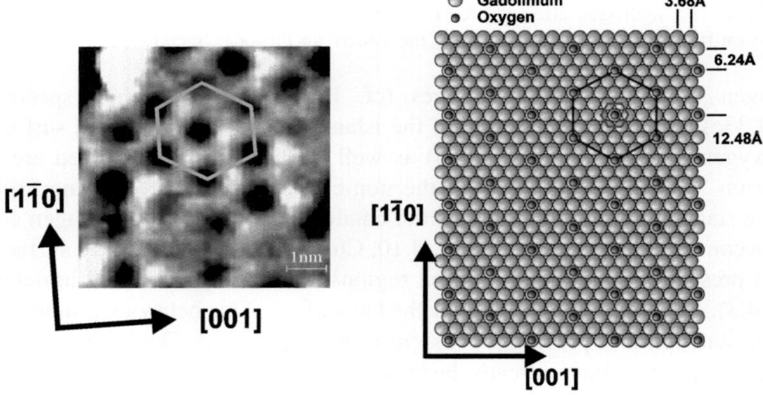

Fig. 4.35 *Left* the same sample as in Fig. 4.32 with an expanded scale. Adsorbed oxygen is imaged as a depression. The adsorption sites of the oxygen atoms form a hexagonal structure which is schematically included with nearest neighbor distances of 13 ± 1 Å. The crystallographic axes are in reference to the tungsten substrate. *Right* the Gd monolayer possessing a (7×14) unit cell is drawn as *light gray balls*. The proposed model of the oxygen overlayer is included by *dark gray balls*. This structure can be attributed to $(2\sqrt{3} \times 2\sqrt{3})R30°$ with a nearest neighbor distance of 12.48 Å. Reprinted with permission from [3]. Copyright (1999) by the American Physical Society

The monolayer of Gd which possesses a rectangular (7×14) unit cell [46] is drawn in the right part of Fig. 4.35 as light gray balls. The interatomic distances were taken from [46]. The proposed model of the adsorbate overlayer is included in this figure which can be attributed to a $(2\sqrt{3} \times 2\sqrt{3})R30°$ structure with a nearest neighbor distance of 12.48 Å being in excellent agreement with the experimental results. The basic unit cells are shown in dark gray color for the Gd monolayer and in black for the adsorbate system, respectively. It should be noted that the presentation with on-top sites is arbitrary and that it is not possible to distinguish between on-top, three-fold hollow, and bridge sites based on the STM observations.

References

1. C. Westphal, J. Bansmann, M. Getzlaff, G. Schönhense, Phys. Rev. Lett. **63**, 151 (1989)
2. M. Getzlaff, M. Bode, R. Wiesendanger, Phys. Rev. B **58**, 9681 (1998)
3. M. Getzlaff, M. Bode, R. Pascal, R. Wiesendanger, Phys. Rev. B **59**, 8195 (1999)
4. B. Kim, A. Andrews, J. Erskine, K. Kim, B. Harmon, Phys. Rev. Lett. **68**, 1931 (1992)
5. D. Li, J. Zhang, P. Dowben, M. Onellion, Phys. Rev. B **48**, 5612 (1993)
6. S. Goldberg, C. Fadley, S. Kono, J. Electr. Spectr. Relat. Phen. **21**, 285 (1981)
7. M. Getzlaff, M. Bode, R. Wiesendanger, Surf. Sci. **410**, 189 (1998)
8. J. Gadzuk, Phys. Rev. B **12**, 5608 (1975)
9. E. Condon, G. Shortley, *The Theory of Atomic Spectra* (University Press cambridge, Cambridge, 1967)
10. K. Horn, A. Bradshaw, K. Jacobi, Surf. Sci. **72**, 719 (1978)
11. M. Scheffler, K. Kambe, F. Forstmann, Sol. State Comm. **25**, 93 (1978)
12. M. Getzlaff, M. Bode, S. Heinze, R. Pascal, R. Wiesendanger, J. Magn. Magn. Mater. **184**, 155 (1998)
13. D. Weller, S. Alvarado, J. Appl. Phys. **59**, 2908 (1986)
14. M. Getzlaff, M. Bode, R. Pascal, R. Wiesendanger, Appl. Surf. Sci. **142**, 63 (1999)
15. R. White, M. Lee, N. Tucker, S. Barrett, P. Murray, Phys. Rev. B **56**, 10071 (1997)
16. D. Menzel, J. Fuggle, Surf. Sci. **74**, 321 (1978)
17. M. Bode, M. Getzlaff, S. Heinze, R. Pascal, R. Wiesendanger, Appl. Phys. A **66**, S121 (1998)
18. K. Ray, X. Pan, E. Plummer, Surf. Sci. **285**, 66 (1939)
19. A. Pundt, M. Getzlaff, M. Bode, R. Kirchheim, R. Wiesendanger, Phys. Rev. B **61**, 9964 (2000)
20. G. Song, A. Remhof, K. Theis-Bröhl, H. Zabel, Phys. Rev. Lett. **79**, 5062 (1997)
21. Q. Yang, G. Schmitz, S. Fähler, H. Krebs, R. Kirchheim, Phys. Rev. B **54**, 9131 (1996)
22. U. Laudahn, A. Pundt, M. Bicker, U.v. Hülsen, U. Geyer, T. Wagner, R. Kirchheim, J. Alloys Comp. **293–295**, 490 (1999)
23. U. Laudahn, Ph.D. thesis, Universität Göttingen (1998)
24. T. Schober, Scripta Met. **7**, 1119 (1973)
25. W. Nix, Metall. Trans. A **20**, 2217 (1989)
26. Y. Fukai, *The Metal-Hydrogen System* (Springer, Berlin, 1993)
27. M. Getzlaff, J. Bansmann, C. Westphal, G. Schönhense, J. Magn. Magn. Mater. **104–107**, 1781 (1992)
28. M. Getzlaff, J. Bansmann, G. Schönhense, J. Chem. Phys. **103**, 6691 (1995)
29. C. Searle, R. Blyth, R. White, N. Tucker, M. Lee, S. Barrett, J. Synchrotron Radiat. **2**, 312 (1995)
30. T. Rhodin, C. Brucker, Sol. State Comm. **23**, 275 (1977)
31. M. Getzlaff, J. Bansmann, G. Schönhense, Fresenius J. Anal. Chem. **353**, 748 (1995)
32. M. Getzlaff, M. Bode, R. Wiesendanger, Appl. Surf. Sci. **142**, 428 (1999)
33. T. Gouder, C. Colmenares, J. Naegele, J. Spirlet, J. Verbist, Surf. Sci. **264**, 354 (1995)
34. K. Wandelt, C. Brundle, Surf. Sci. **157**, 162 (1985)
35. J. Zhang, P. Dowben, D. Li, M. Onellion, Surf. Sci. **329**, 177 (1995)

36. R. Blyth, C. Searle, N. Tucker, R. White, T. Johal, J. Thompson, S. Barrett, Phys. Rev. B **68**, 205404 (2003)
37. M. Budke, J. Correa, M. Donath, Phys. Rev. B **77**, 161401 (2008)
38. M. Budke, M. Donath, Phys. Rev. B **79**, 075432 (2009)
39. F. Netzer, E. Bertel, in: Gschneidner, K., Eyring, L. (eds.) *Handbook on the Physics and Chemistry of Rare Earths*, vol. 5 (North Holland, Amsterdam, 1982)
40. M. Getzlaff, J. Paul, J. Bansmann, C. Ostertag, G. Fecher, G. Schönhense, Surf. Sci. **352–254**, 123 (1996)
41. R. Dubs, S. Dixit, V. McKoy, Phys. Rev. Lett. **54**, 1249 (1985)
42. G. Fecher, Europhys. Lett. **29**, 605 (1995)
43. D. Weller, D. Sarma, Surf. Sci. **171**, L425 (1986)
44. K. Johnson, R. Wilson, S. Chiang, Phys. Rev. Lett. **71**, 1055 (1993)
45. N. Lang, Comments Condens. Matter Phys. **14**, 253 (1989)
46. E. Tober, R. Ynzunza, C. Westphal, C. Fadley, Phys. Rev. B **53**, 5444 (1996)

Chapter 5
Magnetic Characterization

Information on the spin resolved band structure of ferromagnetic materials can *directly* be obtained from spin resolving photoelectron spectroscopy. Using polarized radiation spin integrating photoemission techniques already enable to have access to magnetic properties. An enhancement of the surface sensitivity can be achieved using neutral excited spin polarized atoms which move towards the sample and are de-excited by tunneling electrons from the surface with a subsequent emission of electrons.

In the first part of this chapter it will be reported on spin dependent transport and surface magnetic properties of itinerant magnetic substrates, thin Fe(110) and Co(0001) films evaporated on W(110), which were investigated by these electron emission techniques. Subsequently, the behavior of adsorbates will be discussed from the point of view whether they change the properties of the surface and whether they "feel" the magnetism of the underlying substrate. This discussion will be carried out for the example of oxygen which adsorbs dissociatively on the above mentioned surfaces.

The spectroscopic capabilities of the scanning tunneling microscope [1] open up the fascinating possibility of correlating the local structural and electronic properties with magnetic ones on the atomic scale. Thus, in the second part of this chapter spin polarized tunneling will be demonstrated by measuring the asymmetry of the differential tunneling conductance at bias voltages corresponding to the energetic positions of the two spin contributions of an exchange split surface state in an external magnetic field. This enables the electronic and magnetic structure information to be clearly separated. By mapping the spatial variation of the asymmetry parameter it is possible to observe the nanomagnetic domain structure of Gd(0001) ultra thin films with a lateral resolution on the nanometer scale.

A comprehensive overview on the fundamentals of magnetism and, especially, the magnetic behavior of low-dimensional systems can be found in [2].

M. Getzlaff, *Surface Magnetism*, Springer Tracts in Modern Physics, 240,
DOI: 10.1007/978-3-642-14189-8_5, © Springer-Verlag Berlin Heidelberg 2010

5.1 Itinerant Ferromagnetic Materials

In the following spin dependent transport properties for Co(0001) and surface magnetic properties for epitaxially grown *flat* Fe films on W(110) will be discussed. Then a comparison will be drawn to *nanostructured* iron and cobalt systems. The adsorption of non-magnetic atoms on the surface alters the magnetic properties of the ferromagnetic substrate. These changes will be discussed for oxygen on Fe(110) and Co(0001). Additionally, the question whether the adsorbate atom "feels" the ferromagnetism will be answered.

5.1.1 Spin Dependent Transport Properties

A suitable tool for measuring the attenuation length λ for electrons in solids is the investigation of the intensity of electrons emitted from the substrate while varying the thickness d of an overlayer which is deposited on the concerning substrate surface. A subsequent fit with an exponential decay $e^{-d/\lambda}$ gives the possibility to determine the value of λ. Neglecting elastic scattering effects, the attenuation length describes the inelastic mean free path (IMFP) [3]. The IMFP shows a minimum at electron energies of about 10–100 eV, depending on the material, with a monotonic increase towards lower and higher energies [4].

The knowledge of λ is of interest for surface sensitive experiments, especially the spin dependence of λ for investigations of ferromagnetic substrates. Additionally, it is of great importance in measurements and applications which make use of *spin dependent transport* properties like the giant magnetoresistance (GMR). The spin polarization of secondary electrons is dominated by the magnetism of the topmost few layers. Its determination offers the possibility to investigate surface magnetism and its alterations by roughening on an atomic level induced for example by sputtering or by adsorbates [5]. The spin dependence of IMFP for ferromagnetic materials was first indirectly concluded from the enhancement in the polarization of secondary electrons [6] and from spin polarized electron energy-loss spectroscopy (SPEELS) experiments [7]; a direct proof was reported by Pappas et al. [8] for thin fcc Fe films on Cu(100).

Spin resolving photoemission spectroscopy directly yields the majority and minority spin energy distribution curves thus allowing to distinguish between substrate and overlayer features. Spin resolved spectra are presented in Fig. 5.1 [9] for different thicknesses of Co on W(110). The tungsten related structures at a binding energy of 2.9 eV are shaded, filled upward triangle denote the majority and filled downward triangle the minority spin channel. The different W intensities in the two channels are clearly visible at higher coverages. The peak areas were obtained by integration assuming a linear background indicated by dark gray (for majority) and light gray shaded areas (for minority electrons). The polarization of the electrons emitted from the tungsten substrate depends significantly on the film thickness. Therefore, an induced magnetization of W can be ruled out. In contrast to

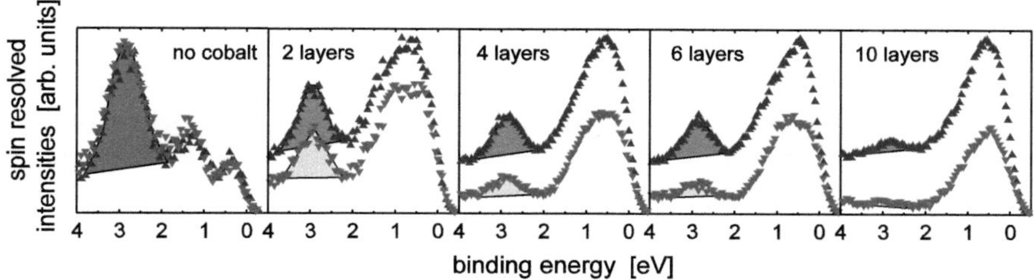

Fig. 5.1 Spin resolved intensities of Co/W(110) after excitation with Ne I radiation ($hv = 16.85\,eV$). *Filled upward triangle* denote majority and *filled downward triangle* minority electrons. The tungsten structure at a binding energy of 2.9 eV shows significantly different peak areas in both spin channels (*light* and *dark* gray shaded) indicating a spin dependence of the particular attenuation lengths

chemisorbed atoms such as oxygen on ferromagnetic substrates [10] an exchange splitting for the tungsten induced structure is not visible. The spin resolved photoelectron intensities I_+ and I_- in the majority and minority spin channel of the photoemission structures of the W(110) substrate crystal can be described as

$$I_\pm = \frac{1}{2} \cdot I_0 \cdot e^{-d/\lambda_\pm} \tag{5.1}$$

with I_0 being the unpolarized intensity of the uncovered tungsten substrate. The spin integrated intensity $I(d)$ is given by

$$I(d) = I_0 \cdot e^{-d/\lambda} \tag{5.2}$$

with $\lambda = 1/2(\lambda_+ + \lambda_-)$ being the spin averaged attenuation length if there are not too large differences between λ_+ and λ_-.

The left part of Fig. 5.2 shows the attenuation of the photoemission signal from tungsten versus film thickness; filled diamond denote the intensity in the spin

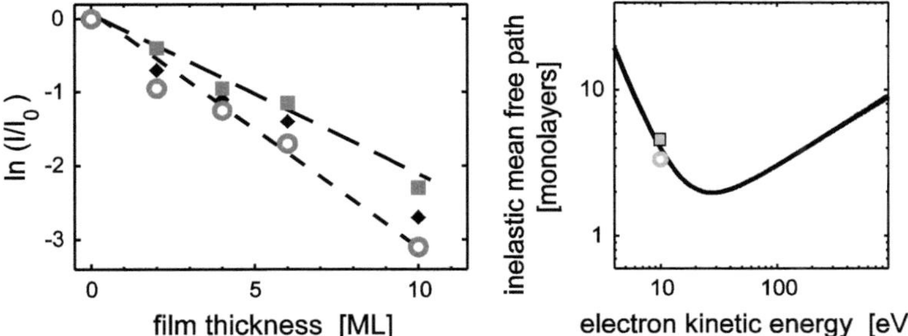

Fig. 5.2 *Left* attenuation of the tungsten feature as a function of cobalt thickness in monolayers (ML) for the spin integrated measurement (*filled diamond*), for majority (*filled square*) and minority (*open circle*) electrons. *Right* universal curve of the inelastic mean free path [4]; included are the different attenuation lengths for majority (*filled square*) and minority (*open circle*) electrons being 4.4 and 3.2 ML, respectively, at a kinetic energy of 10 eV

integrated investigation, filled square the one for majority and open circle for minority electrons. The attenuation lengths are obtained by a least-squares fit, represented by the long- and short-dashed lines. The measured spin averaged IMFP being $\lambda = 3.8$ ML shows good agreement with the "universal curve" (see right part of Fig. 5.2) using the parameters for elements given in [4] leading to about $\lambda = 3.5$ ML. Due to the excellent signal-to-background ratio and the very low statistical scattering the values for the spin dependent IMFP λ_+ and λ_- can directly be obtained from the spin resolved intensities, in contrast to an investigation of Fe/Cu(100) [8]. The kinetic energy of the photoelectrons is about 9.5 eV and the electron energy, referenced to the Fermi level, about 14 eV. Thus, the attenuation lengths amount to $\lambda_+ = 4.4$ and $\lambda_- = 3.2$ ML for the system hcp Co/W(110), respectively.

It seems to be possible to explain the polarization enhancement of secondary electrons by the spin dependence of the IMFP. This means, the different attenuation lengths act as a spin filter, majority electrons preferentially allowing to be transmitted. The spin asymmetry of the IMFP, given by $A = (\lambda_+ - \lambda_-)/(\lambda_+ + \lambda_-)$, amounts to about 20% for both systems and is confirmed by an investigation of Fe/Cu(100) [8] leading to the same value of A. A very effective spin filter can be realized by a graphene layer which was theoretically predicted [11] and experimentally verified for graphene on Ni(111) [12].

The polarization of the secondary electrons can be estimated by

$$P = \frac{I_+ - I_-}{I_+ + I_-} = \frac{e^{-d/\lambda_+} - e^{-d/\lambda_-}}{e^{-d/\lambda_+} + e^{-d/\lambda_-}} \tag{5.3}$$

at a thickness d which the substrate induced structure can just be observed at, in this case $d = 10$ ML, and is given by 40% for hcp cobalt.

5.1.2 Magnetic Properties of Epitaxially Grown Films

With increasing film thickness the electronic structure of the adsorbate overlayer becomes undistinguishable to that of the bulk material; this will be demonstrated for Fe evaporated on W(110) [13]. In the following it will be shown that a thickness of 20 layers is already sufficient to map the transitions in the spin resolved bulk band structure of Fe(110). The analogous behavior was found for Co(0001) films on W(110) [14].

Detecting photoelectrons from Fe(110) in *normal emission* corresponds to transitions along the Γ–Σ–N direction. The spin resolved spectra for the energy region of the valence band are shown in Fig. 5.3 for excitation energies of 21.22 and 16.85 eV; the corresponding transitions are marked in Fig. 5.4. In Fig. 5.3a, the peak in the majority spin channel at a binding energy of 0.7 eV results from the excitation from a $\Sigma_1^{\uparrow}$ and a $\Sigma_4^{\uparrow}$ band, the weak one at 2.4 eV from a $\Sigma_1^{\uparrow}$ band. The corresponding structure in the minority spin channel ($\Sigma_1^{\downarrow}$ band) can be observed at

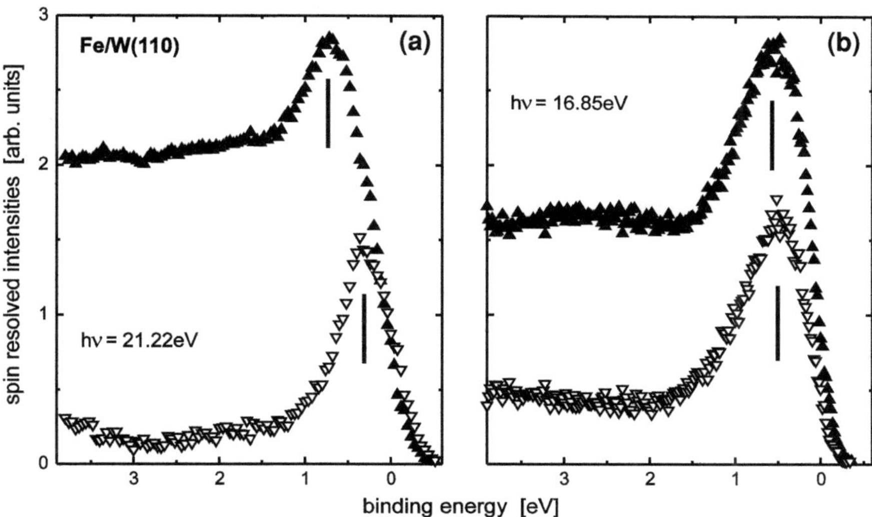

Fig. 5.3 Spin resolved intensities of Fe/W(110) after excitation with **a** He I ($hv = 21.22$ eV) and **b** Ne I ($hv = 16.85$ eV) *light*. The thickness was about 20 layers (from [13], used with permission)

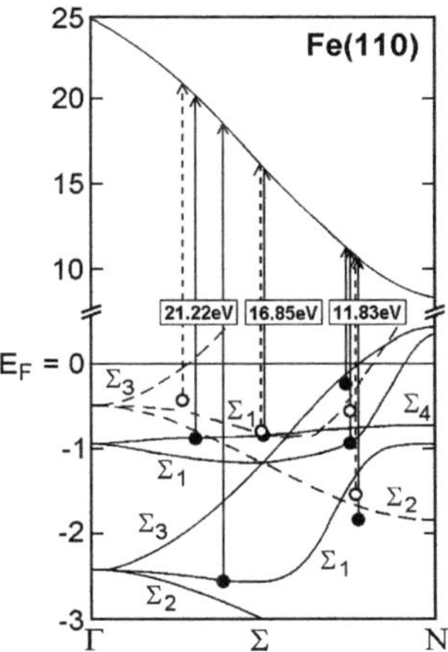

Fig. 5.4 Spin resolved band structure for bcc-Fe along the Σ direction (adapted from [24]). Majority bands are shown as *solid lines*, minority bands as *broken lines*. The *arrows* represent the direct transitions corresponding to the experimentally observed structures in the majority (*filled circle*) and minority (*open circle*) spin channel (cf. Fig. 5.3) (from [13], used with permission)

a binding energy of 0.3 eV, being shifted due to exchange interactions. The resulting value of the exchange splitting amounts to $\Delta E_{ex} = 2.1 \pm 0.2$ eV, being in agreement with measurements of Schröder et al. [15] with about 2 eV as well as of Turner and Erskine [16] with 2.1 eV and also with theoretical band structure calculations of Callaway and Wang [17] as well as of Redinger et al. [18]. The exchange splitting additionally depends on the substrate temperature as could be demonstrated by Kisker et al. for Fe(100) [19, 20].

The magnetic moments for the bulk of $3d$ transition metals amount to 2.2 μ_B, 1.7 μ_B, and 0.6 μ_B for iron, cobalt, and nickel, respectively [21]. These values are reduced in comparison to the corresponding atoms due to hybridization within the solid state. Angle and spin resolving photoelectron spectroscopy is unfortunately unable to determine directly magnetic moments. This determination becomes, however, indirectly possible by comparing experimental spectra with spin resolved band structure and photoemission calculations. The measured exchange splittings of the occupied bands can be compared with the corresponding calculated splitting. Generally for ferromagnetic materials, the exchange splitting of the $3d$ bands near the Fermi edge is approximately proportional to the magnetic moment; every μ_B leads to $\Delta E_{ex} \approx 1$ eV [22]. The direct transitions which correspond to the observed structures are marked in Fig. 5.4. The $E(k_\perp)$ points, determined in this experiment, are shifted by about 0.2 eV towards higher binding energies. This discrepancy of the energy scale points to the influence of correlation effects and the importance of self-energy corrections in the description of the photoemission process.

Changes in the binding energies of electrons excited from dispersing bands can be observed by variations of different experimental parameters. One possibility to determine dispersion effects is the excitation with different photon energies as already done for Fe(110) crystals in a spin integrated measurement [23]. A comparison of the experimental data and theoretical spectra is given in [13]. It is obvious from both experiment and theory that the majority peak position near the Fermi edge shifts only by 0.1 eV when exciting with He I or Ne I radiation. The reason for such a small energetic variation can be found in a nearly vanishing dispersion of the $\Sigma_1^\uparrow$ and $\Sigma_4^\uparrow$ bands near the middle between Γ–N (see Fig. 5.4). The double peak majority structure is resolved in the calculated spectrum for $hv = 21.22$ eV (not shown here; see [13]). The high energy shoulder disperses to higher binding energies for $hv = 16.85$ eV and becomes additionally suppressed. Also in very good agreement between experiment and theory the $\Sigma_1^\downarrow$ peak in the minority spin channel shifts by about 0.2 eV ($E_B = 0.5$ eV) when changing the excitation energy from He I to Ne I. Therefore, it can be concluded that the $\Sigma_1^\downarrow$ band disperses to higher binding energies along the Σ direction.

It was demonstrated that information on magnetic properties can be obtained in a very direct way using spin resolving photoelectron spectroscopy which is unfortunately a distinctly time-consuming technique due to the low efficiency of spin polarization detection systems [25]. Therefore, the question arises whether it is possible to characterize magnetic systems in non-spin resolving, i.e. spin integrated, investigations. This will now be discussed using *circularly polarized* radiation for the excitation process (see Chap. 2.1.3).

In the upper part of Fig. 5.5 photoelectron spectra of a 7 ML thick iron film on W(110) taken with left and right circularly polarized light are displayed for opposite magnetization direction (M^+ and M^-). The required geometric arrangement was presented in Fig. 2.2. The data clearly show a huge intensity difference in the main iron $3d$ band directly at the Fermi level [26], the corresponding asymmetry A:

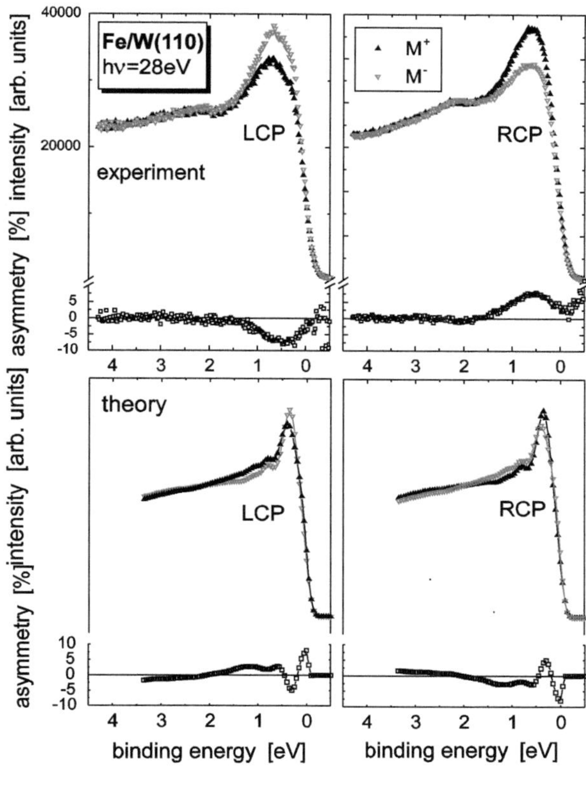

Fig. 5.5 Valence band photoelectron spectra from a 7 ML thick Fe(110) films on tungsten taken with left (*LCP*) and right circularly polarized (*RCP*) radiation ($hv = 28$ eV) for opposite magnetization directions. *Upper panel* experimental data, *lower panel* calculations for Fe(110) with magnetization direction in-plane along the $[1\bar{1}0]$ axis. The corresponding asymmetry is shown in each diagram. Reprinted from [26], Copyright (1998), with permission from Elsevier

$$A = \frac{I^{M+} - I^{M-}}{I^{M+} + I^{M-}} \tag{5.4}$$

reaches values of about $\pm 4\%$ depending on the helicity of the incoming photon beam. This magnetic circular dichroism in the angular distribution of photoelectrons (MCDAD) effect is caused by the combined action of spin–orbit and exchange interaction together with the dipole selection rules for circularly polarized radiation.

The experimental data are confirmed by fully relativistic photoemission calculations carried out by Braun [13] (lower part of Fig. 5.5) showing in general a similar behavior in the photoelectron intensities for both magnetization directions. The calculated data clearly display sharper photoemission peaks than observed in the experiment. Possible damping processes have been considered by adding an imaginary contribution $-iV_{0i}$ to the muffin-tin potentials of Fe. For the initial state this value has been chosen as $V_{0i1} = 0.2$ eV, and the corresponding value for the final state is $V_{0i2} = 1.5$ eV. All calculated spectra have been convoluted with a Fermi function at room temperature and a Gaussian with a full width at half maximum (FWHM) being 0.2 eV to account for the spectrometer resolution. The large fraction of secondary electrons have been simulated by adding 30% of the integrated theoretical intensity between E_F and E to the calculation. The maximum of the calculated feature is shifted by about 0.3 eV towards the Fermi level. This deviation, as already mentioned above, can be attributed to strong correlation effects in the $3d$ bands of ferromagnetic Fe and shows the importance of many-body corrections in the description of ferromagnetic $3d$ metals. Nevertheless, the data exhibit different intensities in the main peak giving rise to an asymmetry with values of twice the experimental data.

The same experiment was carried out using *linearly* polarized light. The photoemission spectra [27, 28] also show differences in the intensities for both magnetization states. The corresponding asymmetries were determined as a function of excitation energy. These results are presented in Fig. 5.6. Whereas for circularly polarized light the asymmetry increases with higher photon energies followed by a decrease the asymmetry for linearly p polarized radiation is much more pronounced [for s polarized light no magnetic linear dichroism in the angular distribution of photoelectrons (MLDAD) effect was observed]. After a minimum at 17 eV with $A_{\mathrm{MLDAD}} = -7\%$ the asymmetry vanishes at about 23 eV. A further deep minimum occurs at 28 eV with -12%. A theoretically calculated band structure [29] shows a large band gap between 25 and 35 eV with only two flat bands above 30 eV.

By comparing the MCDAD results with the band structure one can clearly see that the highest asymmetry values are directly connected to this energy range where no final states from the free electron parabola are available. In contrast to MCDAD the MLDAD asymmetry curve shows a completely different energetic dependence. The minima located at 17 and 28 eV cannot be attributed to the band gap between 25 and 35 eV; instead, they are related to the energy region with high dispersion in the final state bands. This indicates that the information which can be obtained from MCDAD and MLDAD investigations is of complementary nature.

Fig. 5.6 Energy dependence of the MCDAD asymmetry (*top*) and MLDAD asymmetry (*bottom*) for an Fe(110) film on W(110). Reprinted from [27], Copyright (1996), with permission from Elsevier

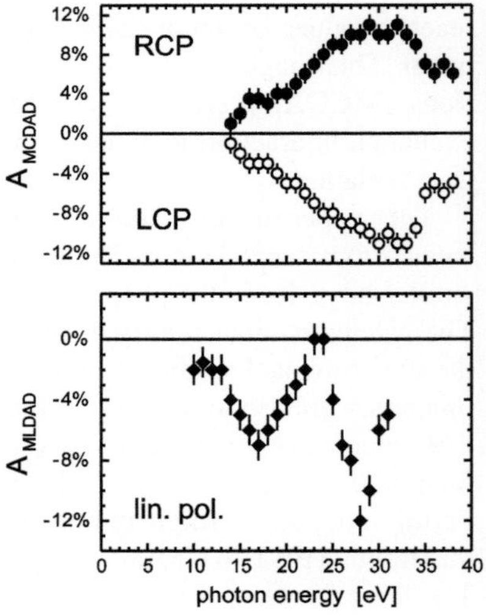

The results presented above were obtained in normal emission. Now, the effect of changing the parallel component of the electron momentum vector, i.e. $k_\parallel$ dispersion, will be discussed.

Figure 5.7 shows spin resolved photoelectron data (left part) and MCDAD spectra (middle LCP, right: RCP) taken at $hv = 21.2$ eV for emission angles from $0°$ to $15°$. In the spin polarized ultraviolet photoelectron spectroscopy (SPUPS) data one can easily recognize huge intensity differences in both spin channels and a dispersion of the minority band (open symbols) to higher binding energies with increasing emission angle, whereas the position of the majority band is nearly unchanged. Hence, the excitation in the minority channel can be attributed to a dispersing band [13]. The MCDAD results show different intensities only in the main iron $3d$ band directly at the Fermi level. These data also display the dispersion of the $3d$ band to higher binding energies. In contrast to the spin

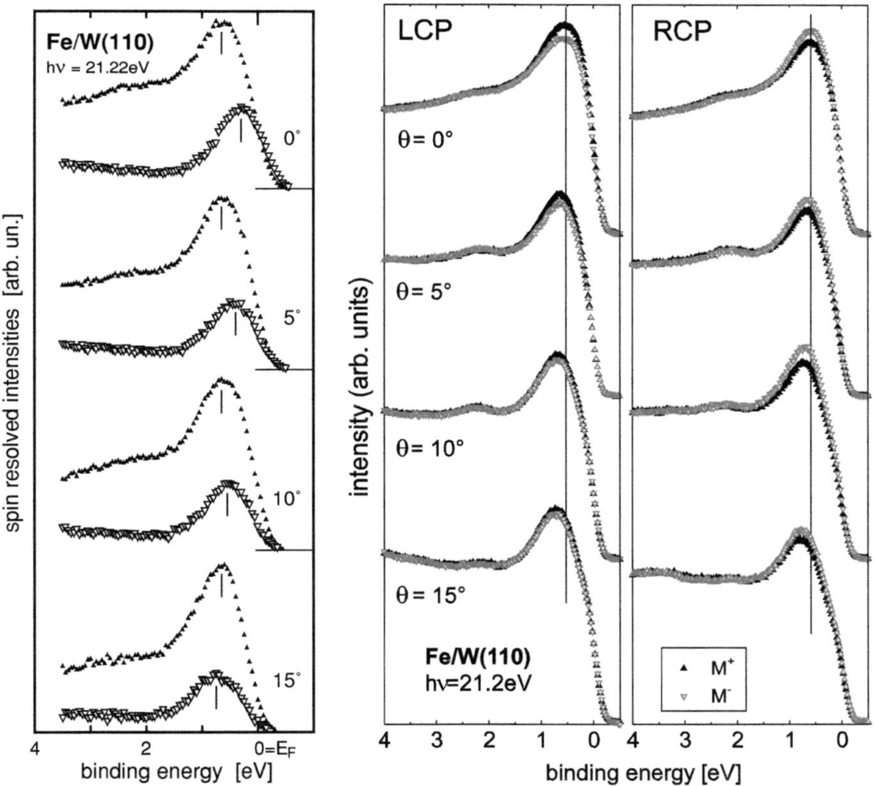

Fig. 5.7 *Left* spin resolved photoemission spectra for thin iron films on W(110) taken with unpolarized light ($hv = 21.22$ eV). *Right* spin integrated spectra taken with circularly polarized radiation ($hv = 21.2$ eV) for emission angles between $0° \leq \theta \leq 15°$. The *thin lines* in the MCDAD spectra mark the position of the valence band maxima in normal emission. Reprinted from [26], Copyright (1998), with permission from Elsevier

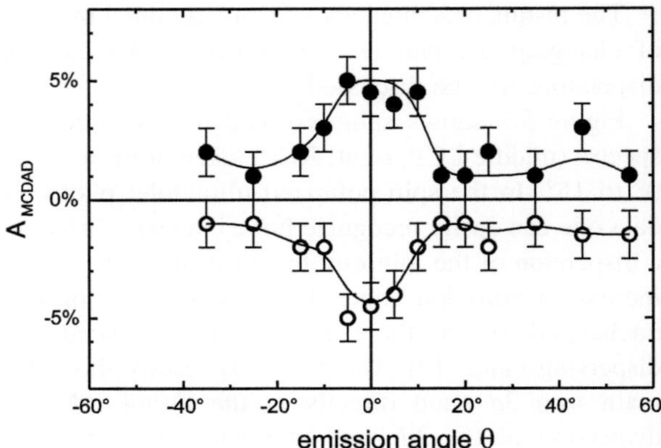

Fig. 5.8 Maximum asymmetry A_{MCDAD} obtained from photoemission spectra taken with left and right circularly polarized light ($hv = 21.2$ eV) as a function of the emission angle θ. Full symbols (*filled circle*) denote MCDAD asymmetry values for RCP, open symbols (*open circle*) for LCP radiation. The *thin solid curves* are to guide the eye. Reprinted from [26], Copyright (1998), with permission from Elsevier

resolved data, both photoemission curves (i.e., for magnetization M^+ as well as M^-) exhibit the same shift. Furthermore, the MCDAD asymmetry becomes definitively smaller with increasing emission angle, which can also be seen in Fig. 5.8 showing a larger range of the emission angle. The MCDAD effect is still present (typically $\pm 2\%$) for emission angles of more than $40°$. This observation points to a different origin of the asymmetries. Whereas in spin resolved photoemission the different photoelectron intensities are due to the exchange splitting of the occupied bands, in MCDAD experiments one probes the general shape of the bands like, for example, hybridization points in k space.

As discussed above magnetic dichroism in the angular distribution of photoelectrons (MDAD) from ferromagnetic materials can be observed using circularly polarized (MCDAD) [30] and linearly polarized (MLDAD) light for electrons in the valence band near the Fermi level [31]. A theoretical consideration [32] shows that the MDAD effect from core level electrons is also present and can successfully be described using the pure atomic model. From this theory it follows that the analogous effect should be observable even using unpolarized light. The same conclusion was drawn in [33], while in other theoretical approaches [34, 35] only absorption of circularly and linearly polarized light was discussed.

In the following it will be demonstrated that magnetic dichroitic effects not necessarily need linearly or circularly polarized light but can already be observed with *unpolarized* light (being named UMDAD).

In view of the former belief that dichroitic effects in magnetic systems require the use of circularly or linearly polarized light, these results are surprising. In several respects, the situation can be compared to the analogous case of photoelectron spin polarization for unoriented systems (e.g., free atoms). After the discovery of the Fano effect (polarized electrons from non-magnetic atoms produced by photoemission with circularly polarized light [36, 37]) it was theoretically predicted [38] and later experimentally confirmed that also unpolarized [39] and linearly polarized light [40] leads to the emission of polarized electrons from free atoms. It was crucial for the latter phenomena that the fine structure splitting of the bound state was resolved otherwise the polarizations of the two spin–orbit components would cancel each other. The experimental arrangements for observing the spin phenomena are characterized by three vectors, the vector of the photon polarization (or the direction of the photon beam for unpolarized light), the photoelectron momentum vector, and the vector of the photoelectron spin. In the case of MDAD from ferromagnetic materials, the first two vectors remain the same while the third one is substituted by the direction of the initial spin polarization (i.e. magnetization) of the atom.

The crucial point for observing MDAD is to resolve the photoelectrons from different magnetic sublevels. The spin–orbit interaction does not play the key role in producing MDAD since the parameters describing MDAD are analogous to the angular asymmetry parameter β [41] which is approximately the same for the two spin–orbit components. Even in absence of the spin–orbit interaction, MDAD should be observed provided the splitting of sublevels with different projections of the orbital angular momentum m is resolved. The spin–orbit splitting in ferromagnetic materials is at least of the same order of magnitude as the magnetic splitting, and the magnetic sublevels are defined by projections m_j of the total angular momentum j of the subshell. Similar to the spin polarization of photoelectrons ejected from unpolarized atoms, MDAD from ferromagnetic materials should exist for absorption of circularly polarized, linearly polarized, and unpolarized light.

In this experiment [42], the photoelectrons are detected in the direction of the surface normal. The beam of unpolarized light forms an angle θ_{ph} with the surface normal. The direction of the sample magnetization $\mathbf{M}$ is perpendicular to both the surface normal and the photon beam. It follows from theory [32] that for this geometry and for the absorption of *linearly* polarized light with polarization perpendicular to $\mathbf{M}$, MLDAD is defined by the equation:

$$I^j_{\mathrm{MLDAD}} = I^j_{\mathrm{M\uparrow}} - I^j_{\downarrow} = \frac{\sigma_{nlj}(h\nu)}{2\pi} W^j \cdot \frac{1}{2}\sin(2\theta_{\mathrm{ph}}) \tag{5.5}$$

with $\sigma_{nlj}(h\nu)$ being the partial photoionization cross section of the nlj subshell, $h\nu$ the photon energy, and

$$W^j = 3i(2j+1)\left[\rho^n_{10}C^j_{221} + \frac{1}{2}\rho^n_{30}\left(C^j_{223} - \frac{\sqrt{10}}{2}C^j_{243}\right)\right] \tag{5.6}$$

ρ^n_{N0} are the state multipoles characterizing the polarization of the hole state [43], and C^j_{kLN} are the parameters defined in [41].

The MLDAD vanishes in normal emission if s polarized light is used keeping the other parameters fixed. Unpolarized light can be described by an incoherent superposition of s and p polarized light. Therefore, for the absorption of *unpolarized* light the effect is two times smaller, but it is described by exactly the same parameters

$$I^j_{\mathrm{UMDAD}} = \frac{\sigma_{nlj}(h\nu)}{2\pi} W^j \cdot \frac{1}{4}\sin(2\theta_{\mathrm{ph}}) \tag{5.7}$$

Apart from the numerical factor, MLDAD and UMDAD are *identical*. It should be noted that $I_{MDAD}(\theta_{ph}) = -I_{MDAD}(-\theta_{ph})$, as can directly be seen from Eqs. 5.5 and 5.7. $I_{MDAD}(\theta_{ph})$ is largest for light incidence of $\theta_{ph} = \pm 45°$.

In the case of valence band photoemission, the atomic model cannot directly be applied to the numerical estimation of the effect, but rather for a qualitative consideration only. For valence bands, the initial state is no longer described by a single spinor spherical harmonic as it was done in [32] but it can be expanded for a certain **k** value in a series of spherical harmonics [44] due to their completeness. This procedure will influence the values of the state multipoles and the dipole matrix elements in Eq. 5.6, but the general Eqs. 5.5 and 5.7 will remain unchanged. In particular, they should correctly describe the dependence of MDAD on the angle of photon incidence.

Figure 5.9 shows the photoelectron energy distribution of Fe/W(110) obtained in normal emission for oppositely magnetized (M^+, M^-) films and the corresponding asymmetry $A = (I^{M+} - I^{M-})/(I^{M+} + I^{M-})$. The angle of photon incidence $(hv = 21.22\ eV)$ was $\theta_{ph} = -30°$. It is obvious that for the iron structures the intensities are *different* thus giving evidence of a *magnetic dichroism in the valence band region*. The highest asymmetry occurs at 0.5 eV with a value of about +10%. Both positive values therefore seem to be related to the emission mainly from bands with majority character.

The information depth in photoemission experiments described above amounts to a few atomic layers due to the kinetic energies of the photoelectrons being about 10 eV [4]. Therefore, the whole discussion was almost related to surface magnetic behavior.

In the following the magnetic properties of the topmost layer will exclusively be discussed by using spin polarized metastable de-excitation spectroscopy [45]. An overview of this experimental technique was given in Chap. 2.2.

Above metallic surfaces the de-excitation of the metastable $He(2^3S)$ atoms takes place via a resonance ionization (RI) with a subsequent Auger neutralization (AN) as described in Chap. 2.2. The effective energy of the $1s$ level concerning He atoms

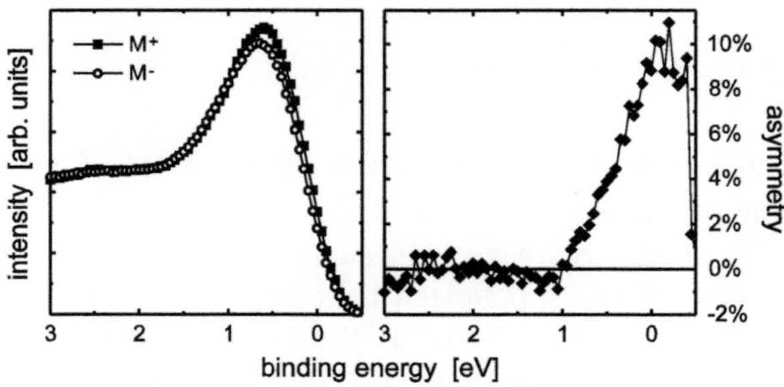

Fig. 5.9 Photoelectron intensities (*left*) and corresponding asymmetry (*right*) of an Fe(110) film on W(110) with a thickness of about 25 Å. The spectra have been taken in normal emission for oppositely magnetized samples and with *unpolarized* light. Reprinted with permission from [42]. Copyright (1994) by the American Physical Society

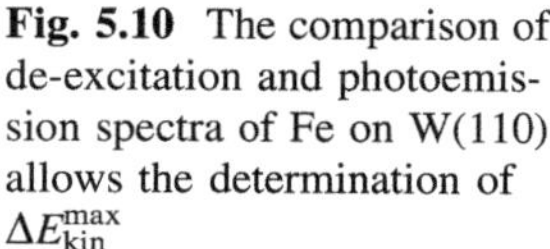

Fig. 5.10 The comparison of de-excitation and photoemission spectra of Fe on W(110) allows the determination of $\Delta E_{\text{kin}}^{\text{max}}$

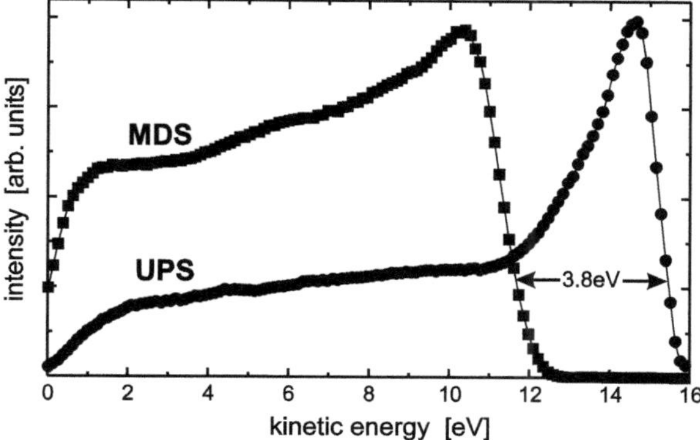

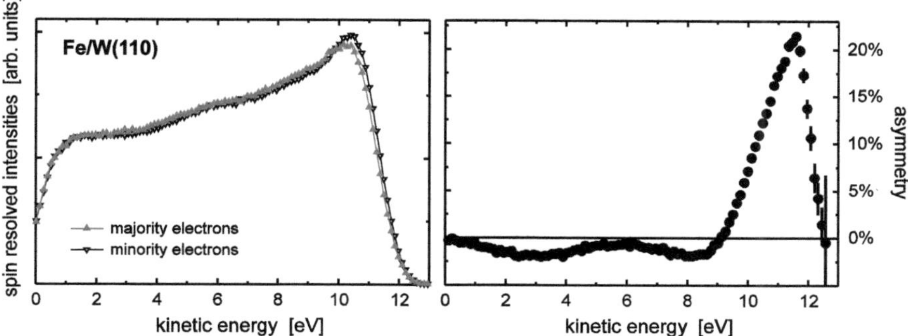

Fig. 5.11 Spin resolved de-excitation spectra (*left*) and corresponding asymmetry (*right*) for Fe/W(110)

E_{1s}^{eff} above a surface is different to that of a free atom E_{1s}. Due to the image potential the effective energy depends on the distance between atom and surface. The different kinetic energies of a common feature in de-excitation and photoemission spectra allows the determination of the effective energy E_{1s}^{eff}. This is demonstrated in Fig. 5.10 for Fe on W(110). The value of $\Delta E_{\text{kin}}^{\text{max}}$ amounts to 3.8 eV (cf. Eq. 2.3). Using $\Phi = 4.8$ eV [46] results in $E_{1s}^{\text{eff}} = 22.2$ eV. The difference to the binding energy in a free atom being $E_{1s} = 24.6$ of 2.4 eV is comparable to He atoms above Pd(111) surfaces (2.3 eV) and polycrystalline tungsten (2.2 eV) [47].

Figure 5.11 shows the spin resolved de-excitation spectra (left) of Fe/W(110) and the corresponding asymmetry (right) as a function of the kinetic energy. The feature between 9 and 11 eV is due to the Fe $3d$ bands. The positive sign of the asymmetry in the high kinetic energy region, corresponding to electrons near the Fermi level, points to a surplus of minority electrons at the near-surface region. This observation is in agreement to investigations of Fe/GaAs(110) [48] and of Fe/Ag(001) [49].

The sign of the asymmetry corroborates theoretical calculations [50] for Fe(110) using the (FLAPW) method. The spin resolved density of states was

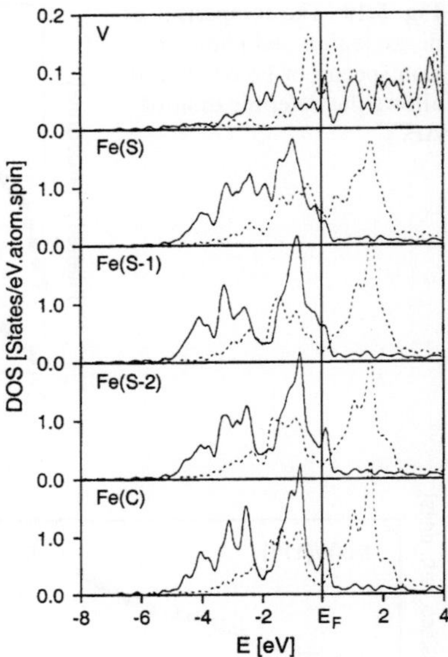

Fig. 5.12 Layer projected density of states including the near-surface vacuum region (V), Fe(S) denotes the outermost iron layer, Fe(C) the center layer of a seven layer thick film (*solid lines* majority electrons; *broken lines* minority electrons). Reprinted with permission from [50]. Copyright (1992) by the American Physical Society

determined for each layer of a seven layer thick iron film starting with the center layer Fe(C) to the surface layer Fe(S) and the near-surface vacuum region V (see Fig. 5.12). The inner layers [Fe(C) and Fe(S − 1)] possess a surplus of majority electrons in the energy region near the Fermi level E_F (see Fig. 5.13). In the outermost layer, however, exhibiting a reduced coordination number a narrowing

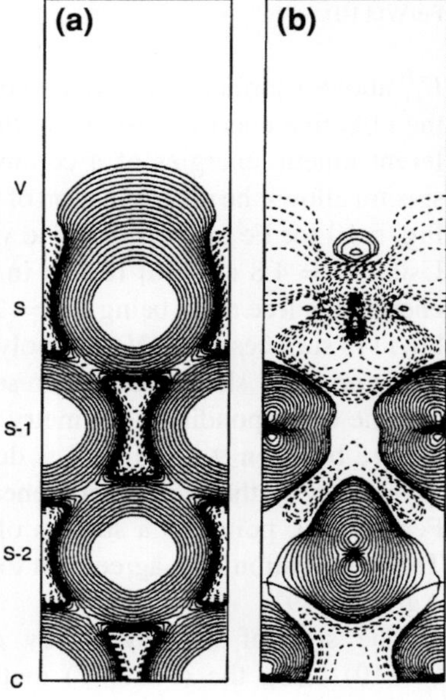

Fig. 5.13 Distribution of the **a** total spin density $m(\mathbf{r})$ and **b** of states near the Fermi level $m(\mathbf{r}, E_F)$ for an Fe(110) layer (cf. Fig. 5.12). The *solid lines* denote majority electrons, the *broken lines* minority electrons. Reprinted with permission from [50]. Copyright (1992) by the American Physical Society

of the d band occurs [51]. The resulting band shift leads to a surplus of minority electrons near E_F at the surface Fe(S) and in the vacuum region V (see Fig. 5.13b). This theoretical result is in good agreement with the positive asymmetry (corresponding to a surplus of minority electrons) at the high kinetic energy side in the de-excitation spectrum for Fe(110) (see Fig. 5.11).

Apparently in contradiction the photoemission spectra (see Fig. 5.3) exhibit a dominance of majority electrons near the Fermi level. However, keeping in mind the calculation of Wu et al. [50] one can now easily realize the *distinct surface sensitivity of metastable de-excitation spectroscopy* (MDS) which gives predominantly information from the topmost surface layer whereas in photoemission experiments the information depth is a few layers.

5.1.3 Nanostructured Systems

The description of magnetic properties up to now referred to epitaxially grown flat films. In the following these results will be compared with nanostructured systems [52–55].

Heating the thin film above a critical temperature results in an island formation [56]. Due to the reduced dimensionality these islands are expected to exhibit different magnetic properties compared to flat films. To obtain magnetic information the MCDAD and MLDAD experiments were extended to the core levels of Fe and Co.

In experiments using magnetic circular and linear dichroism in photoemission the observed photoelectron intensities in the Fe and Co $p_{1/2}$, $p_{3/2}$ levels and $3d$ valence bands depend on the magnetization of the samples and therefore create a small intensity difference. Typical MCDAD spectra taken at the Fe $3p$ level for an epitaxial ten monolayers thick iron film (upper part) and an island structure (lower part) on W(110) are shown in Fig. 5.14. The data display the photoelectron intensities for LCP (left panel) and RCP light (right panel) in dependence of the in-plane magnetization states M^+ and M^-. The upper panel (i.e., the spectra of the epitaxial overlayer) clearly exhibits a small energetic shift in the peak maxima whereas the data in the lower panel of Fig. 5.14 are identical within the error bars. In MCDAD photoemission experiments the intensity difference or the energetic shift depend on the helicity of the radiation and changes its sign when the photon helicity is reversed.

As already mentioned, the phenomenon of magnetic circular dichroism in photoemission originates from spin–orbit and exchange interactions in combination with the dipole selection rules. In the atomic model picture, the splitting of the $3p$ level (into sublevels with orbital momentum m) is caused by the electrostatic interaction of the core level with the magnetically polarized valence electrons [57]. The observed intensity differences and the respective asymmetry values in photoemission from the Fe $3p$ levels are small (typically 3%) compared to the large MCDAD and MLDAD asymmetries (up to about 12%) observed in valence band photoemission [27].

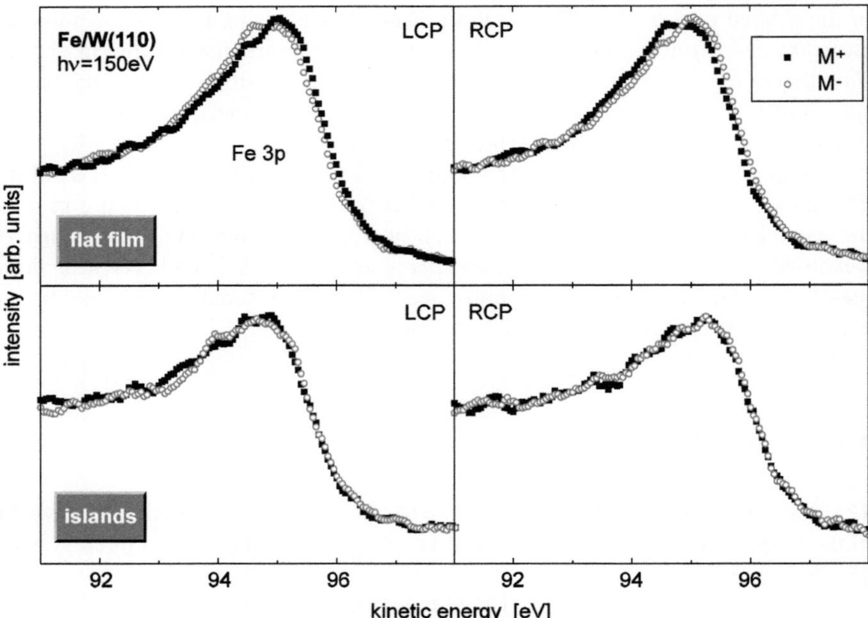

Fig. 5.14 Typical normal emission Fe $3p$ photoemission spectra taken with left and right circularly polarized radiation ($hv = 150$ eV) in dependence of the magnetization M^+ and M^-. *Upper panel* epitaxial 10 ML thick iron overlayer on W(110); *lower panel* iron islands on W(110) after annealing the epitaxial film to about 1,000 K. From [52], used with permission

It is very interesting to observe that the magnetic dichroism vanishes after heating the sample to about 1,000 K. A possible explanation could be that there is no longer a long range ferromagnetic order in the island structure or that the easy magnetization axis of the Fe islands has switched from the $[1\bar{1}0]$ direction for thin Fe films to the [100] direction related to the tungsten substrate. The last possibility is more likely because the iron bcc(110) islands are oriented along the [100] direction. This has been observed in low-energy electron diffraction (LEED) [58] and scanning tunneling microscopy [59]. Now the iron bulk easy magnetization axis (the Fe [100] direction) is parallel to the [100] direction of the W(110) crystal. This results in a vanishing asymmetry due to the lack of chirality in the geometrical arrangement of the experiment (cf. Fig. 2.2).

In contrast to iron on tungsten, the MCDAD data for cobalt island structures on W(110) clearly show a remnant magnetization (see Fig. 5.15). Analogously to Fig. 5.14 the photoemission spectra are displayed for both magnetization states (M^+ and M^-). The left as well as the right panel show different photoelectron intensities in the Co $3p$ levels. In both cases the observed intensity difference is small but again the MCDAD effect changes its sign when the photon helicity is reversed. The cobalt islands result from heating an epitaxial 6 ML Co film to temperatures of about 1,000 K. A strong difference in the LEED pattern before

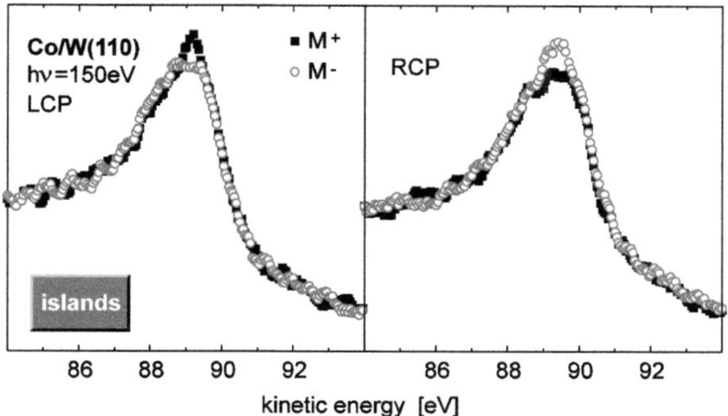

Fig. 5.15 Normal emission photoelectron spectra of a Co island structure on W(110) for the two magnetization states M$^+$ and M$^-$. The spectra were taken with left (*LCP*) and right circularly polarized (*RCP*) synchrotron radiation with a photon energy of $hv = 150$ eV thus probing the MCDAD effect. From [52], used with permission

and after the annealing procedure could be observed. A monolayer of cobalt is located between the island structure and the tungsten substrate. The LEED data showing both a 4×1 superstructure [due to the compressed monolayer on W(110)] and a sixfold symmetry belonging to the hcp structure indicate that this complete cobalt overlayer on tungsten exists.

However, in contrast to the iron island structure on tungsten, where the existence of a complete iron monolayer was shown by STM investigations [59], it was up to now impossible to carry out any element specific analysis in between two islands. Spin resolved photoemission data of Co/W(110) [14] proof that a monolayer of cobalt at room temperature has no remnant magnetization. Therefore, the magnetic effect displayed in Fig. 5.15 cannot result from the cobalt monolayer, the magnetic phenomenon observed here is connected with the island structure.

Figure 5.16 displays photoelectron spectra taken from a 7 ML flat Co film and a cobalt island system (exhibiting an equivalent coverage of 3.5 ML). The data have been recorded in remanence after pulse magnetization in opposite directions (M$^+$ and M$^-$) using linearly polarized light at $hv = 16$ eV. The lower panel shows the corresponding asymmetry. Typical values for the asymmetry are less than 10%, and, of course, the asymmetry varies with the photon energy [26]. Both photoemission data clearly show an MLDAD effect. In the right panel, the intensity difference is much more pronounced for the peak close to the Fermi level E_F. For thin Co films, the MLDAD effect nearly vanishes for this structure but an asymmetry is visible for the peak located at about 1.2 eV binding energy. This observation directly points to the different electronic properties of flat films and islands because the MLDAD effect is caused by a hybridization of bands with different symmetry in the initial state and a non-vanishing exchange splitting in the final state together with the chiral experimental setup. Details for thin cobalt films are described in [26] with respect to a fully-relativistic band structure calculation.

The photoemission peak in the cobalt island spectra located around a binding energy of 3 eV originates from a tungsten state and is only visible at photon energies between 15 and 17 eV. For a 7 ML thick Co film, the intensity is nearly attenuated due to the high surface sensitivity of ultraviolet photoelectron spectroscopy. The photoemission peak at 3 eV binding energy (right part) shows a

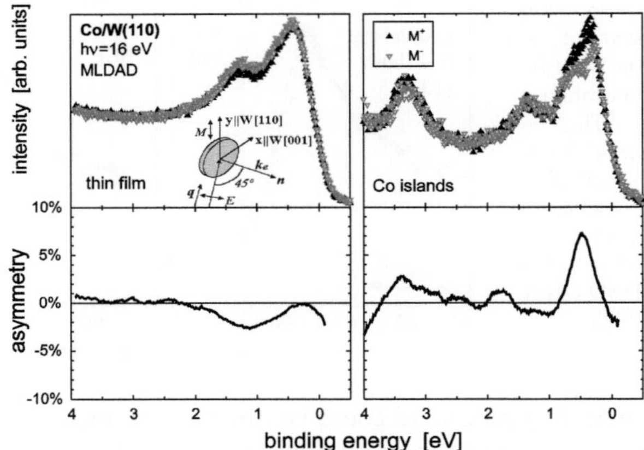

Fig. 5.16 Photoelectron spectra of Co $3d$ valence band for a well-ordered hcp(0001) surface (*left*) and Co islands (*right part*) on W(110) taken for opposite in-plane magnetization directions (*filled triangles* M$^+$, *open triangles* M$^-$) taken with linearly polarized light thus probing the MLDAD effect. In the lower part, the corresponding asymmetry is displayed with the *solid curve* giving an average. The *inset* displays the experimental setup with the induced chirality. The magnetization M has been carried out perpendicular to the orbit plane of the storage ring. From [55], used with permission

small asymmetry even though the tungsten bulk is indeed not magnetically ordered. Similar effects have also been observed in spin resolved photoemission (e.g. Co/W(110) [14]) and in MCDAD (thin iron films on Cu(100) [60]). This phenomenon can be explained by the spin dependent attenuation of electrons (here from the W substrate) which pass through a magnetic layer therefore creating a very effective spin filter (cf. the discussion in Sect. 5.1.1). The inset in Fig. 5.16 shows the arrangement of incoming photon beam **q** (with the linear polarization vector **E** being in the orbit plane), the orientation of the tungsten crystal (surface normal **n**) and the magnetization vector **M**. The photoelectrons are detected in normal emission for thin films or close to normal emission (in the orbit plane less than 10° with respect to **n**) in the case of Co islands.

The energy variation of the photoelectron spectra is displayed in Fig. 5.17 (left part) in the range between 21 and 32 eV, the data have been recorded analogously to Fig. 5.16. For comparison, the results for flat Co(0001) films on W(110) can be found in [61]. In the right part of Fig. 5.17, a band structure of bulk hcp(0001) cobalt is shown which has been fully relativistically calculated for an in-plane magnetization. Since the MLDAD is caused by spin–orbit as well as exchange-splitting, both effects have been taken into account on the same level of accuracy. The electron distribution curves show an energetic dispersion in the prominent feature near E_F. From 21 eV to about 28 eV, the binding energy is nearly constant being 0.75 eV. Above 28 eV, the dispersion starts from 0.75 to 0.4 eV. Additionally, a new feature in the spectra occurs directly at the Fermi level (from 24 eV upwards) and becomes dominant above 30 eV. This peak can be attributed to a

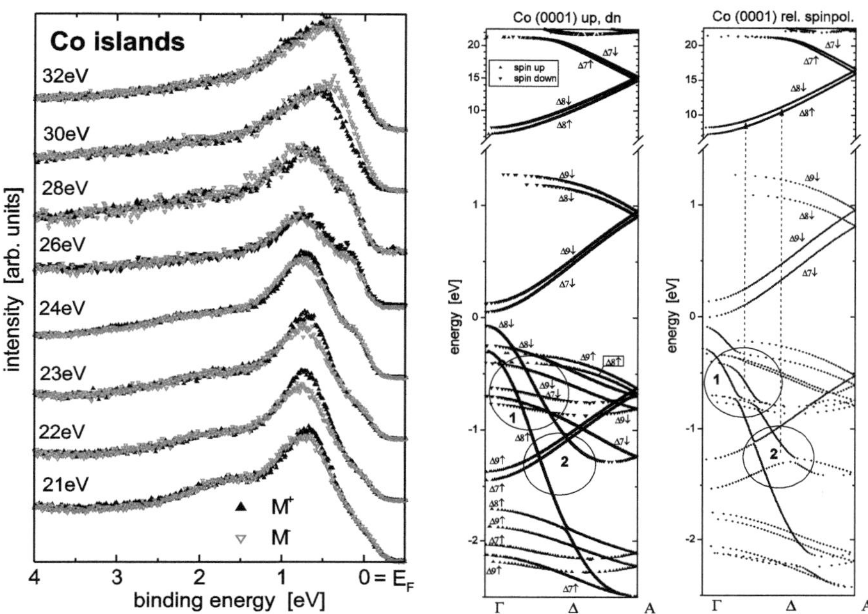

Fig. 5.17 *Left part* energy dependence of the photoelectron spectra analogously obtained to Fig. 5.16. The islands were created by heating an epitaxial cobalt film to about 1,000 K. From [55], used with permission. *Right part* for comparison, a fully relativistically calculated band structure for hcp(0001) bulk cobalt with in-plane magnetization is displayed. This calculation takes into account both spin–orbit interaction and exchange splitting on the same level of accuracy. The *circles* in the initial states show two regions where hybridization effects are present. Reprinted from [28], Copyright (1998), with permission from Elsevier

direct transition from the initial states near the Γ–point (just below E_{F}) into the final state band near 25 eV and disperses to higher binding energies of 0.4 eV. Furthermore, the cobalt band structure has two gaps in the final states as displayed here, the first around 22 eV and the second between 27 and 29 eV.

A possible influence from the underlying tungsten substrate on the experimental data can be excluded, since photoemission spectra of clean tungsten (not shown here) do not exhibit any significant structure in the interesting region between the Fermi level and a binding energy of $E_{\mathrm{B}} = 1.5$ eV. Compared to data from thin epitaxial films with a thickness of about seven monolayers, the photoelectron intensity in the peak structure around the Fermi level (cf. Fig. 5.17, $h\nu = 24$ eV to 28 eV, $E_{\mathrm{B}} \approx 0.25$ eV) is suppressed and below 24 eV the main peak (located at around 0.7 eV) appears at higher binding energies. In principle, photoemission spectra from hcp(0001) Co films with a thickness above 4 ML can be regarded as bulk-like cobalt data [62]. Differences in the photoelectron intensity are mostly discussed with respect to lower coverages, whereas shifts in the binding energy usually arise due to the development from a two-dimensional to a three-dimensional band structure. Here, the energetic shift observed between the Co island data and photoemission data from a 7 ML Co film can, of course, originate due to the confinement of the nanostructured Co islands.

The MLDAD effect shows a maximum (i.e. highest intensity difference) at 22 eV which may be connected with the band gap in this region. Since this gap is quite small and the exchange splitting in the final state above the gap is clearly present, the intensity differences can occur from a preferential excitation for one magnetization direction into these empty states. Furthermore, the peak maxima (for opposite magnetization direction M^+ and M^-) clearly have different energetic positions which are caused by hybridization effects where the energetic bands are not allowed to cross each other (avoided crossover). In the energetic region between 26 and 28 eV no final states of the free electron parabola are available and hence, no significant MLDAD can be observed. A possible reason could be the absence of an exchange splitting in the final state.

Nevertheless, the MLDAD photoemission data shown here clearly demonstrate that 3D cobalt islands with a coverage equivalent to 3.5 ML are ferromagnetically ordered with a remanent magnetization along the $W[1\bar{1}0]$ direction. The magnetic effect for cobalt islands cannot simply result from a possibly existing cobalt interface layer. Spin resolved photoemission data of Co/W(110) [14] as well as magneto-optical Kerr effect (MOKE) measurements [63] prove that at room temperature one monolayer of cobalt has no remanent magnetization. The critical thickness for the onset of ferromagnetism is about 1.6 layers at room temperature, for thinner films, the Curie temperature T_C is reduced. For one cobalt monolayer, T_C is in the region of 260 K. Furthermore, Garreau et al. [63] observed that the Curie temperature of thin cobalt films below three monolayers decreases when annealing the films above 440 K, even though the stress in the adsorbed layer is released. Above a thickness of three ML, on the other hand, T_C increases after annealing. One possible argument for explaining this behavior is that the Co islands created by tempering are large enough to enable a ferromagnetic coupling. The lateral finite size effect should play no important role, the Curie temperature is determined by the heights of the islands. Below this coverage, the Co structure acts as a superparamagnetic system with most likely small and disconnected islands. Therefore, the magnetic phenomenon observed here is connected to the Co island structure.

5.1.4 Influence of Adsorbates

The chemisorption of adsorbate atoms results in changes of the electronic properties at the surface due to hybridization of adsorbate and substrate states. If the substrate is ferromagnetic then the following questions arise:

- Do the magnetic properties of the substrate change?
- Does the non-magnetic adsorbate atom "feel" the ferromagnetism of the underlying substrate?

In the following discussion one will see that the use of SPUPS and spin polarized metastable de-excitation spectroscopy (SPMDS) allows to give the answers due to the capability of a direct access to the adsorbate induced states and the distinct surface sensitivity.

In photoemission spectra the oxygen $2p$ related features can be observed at a binding energy of about 6 eV below the Fermi level (see Fig. 5.18). This structure consists of the p_x level at the low energy and the p_z level at the high energy side. The assignment to p_x and p_z is made possible using linearly polarized radiation or by varying the incident angle [64]. Through the excitation with unpolarized light used in this experiment it is practically impossible to separate both peaks in an unambiguous way. The determination of the peak maxima in each spectrum was carried out by subtracting a linear background and a subsequent fitting of the oxygen induced peak with two Gaussian functions. The O p_x feature on the low binding energy side was well reproducible but the O p_z peak at the high binding energy side was correlated with a large error due to its low intensity; therefore the discussion is restricted to the p_x state which is mainly responsible for the

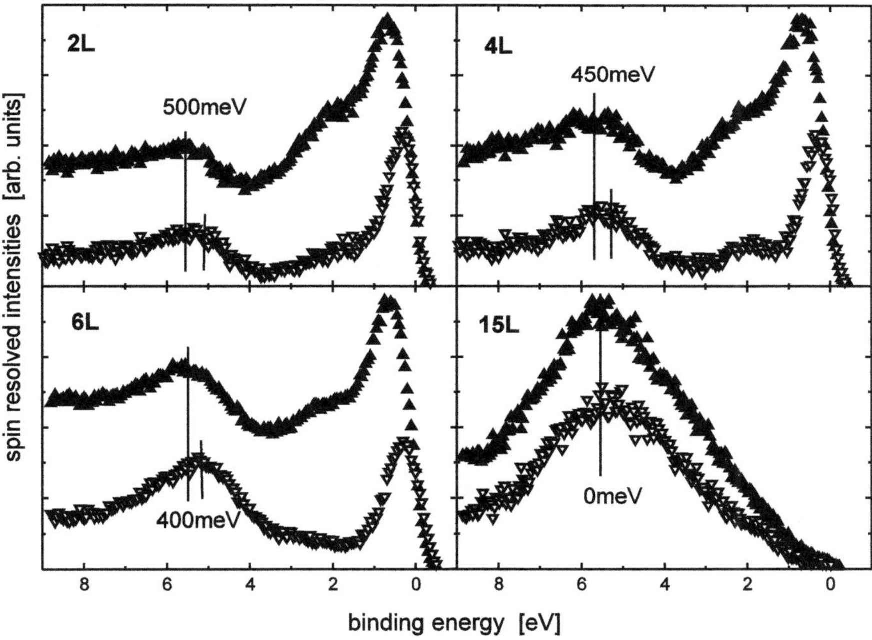

Fig. 5.18 Spin resolved photoemission spectra taken after different exposures of oxygen on Fe/W(110) ($hv = 21.22$ eV). Full symbols (*filled upward triangle*) denote the majority spin channel, open symbols (*open downward triangle*) the minority channel. The films were magnetized in-plane along the $[1\bar{1}0]$ direction of the substrate. The exchange splitting of the O $2p_x$ level of $\Delta E_{\mathrm{ex}} = 500$ meV becomes reduced with increasing exposure and vanishes after 15 L. The shoulder at the high energy side of the oxygen induced structure is caused by the O $2p_z$ state. Reprinted from [65], Copyright (1999), with permission from Elsevier

hybridization with the iron substrate (see [66]). This restriction leads to the fact that the maximum of the peak does *not* coincide with the maximum of the p_x related feature.

An additional oxygen dosage broadens the adsorbate induced peak. The oxidation to FeO starts above an exposure of about 10 L [67]. The related change in the electronic properties becomes apparent from the additional emission at binding energies between 2.5 and 4 eV. A drastic reduction of the iron valence band emission is related with this creation of iron oxide. The spectrum is now dominated by the broad adsorbate induced structure at a binding energy of 5.5 eV. An energetic shift of the oxygen induced peak could not be observed until the beginning of the oxidation process.

Figure 5.18 presents spin resolved spectra in normal emission for different oxygen coverages. Full symbols (filled upward triangle) denote the majority spin channel, open symbols (open downward triangle) the minority channel. The films were magnetized in-plane along the $[1\bar{1}0]$ direction of the iron substrate. Photoelectrons were excited using He I radiation. The exchange splitting of the O $2p_x$ level after an exposure of 2 L amounts to $\Delta E_{ex} = 0.5 \pm 0.1$ eV. With increasing coverage this splitting is reduced to $\Delta E_{ex} = 0.4 \pm 0.1$ eV after 6 L. The structure of the iron valence band near the Fermi edge is very well pronounced in each case and the intensity only slightly reduced. After the transition from chemisorption to the oxidation process and the formation of FeO at the surface following an exposure of about 10 L the exchange splitting has disappeared and the intensity of the Fe $3d$ band is strongly suppressed (see spectrum in Fig. 5.18 after 15 L).

The exchange splitting ΔE_{ex} of the $2p_x$ level of oxygen adsorbed on Fe/W(110) is shown in Fig. 5.19 as a function of oxygen exposure. With increasing coverage the exchange splitting is reduced until it vanishes at the beginning of the oxidation process to antiferromagnetic FeO upon an exposure of about 12 L. In the low exposure regime ($\leq$6 L) the sharp peak around 6 eV and the low intensity at about 2 eV point to exclusively chemisorbed oxygen on the surface. The onset of the broadening at higher exposures ($\geq$10 L) is due to the transformation of the

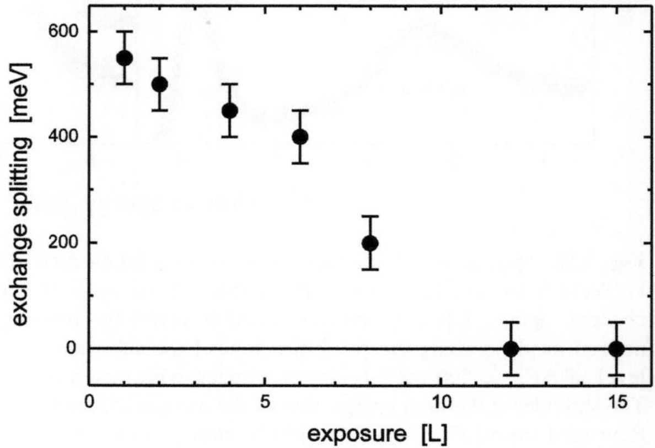

Fig. 5.19 Exchange splitting of the O $2p_x$ level of oxygen being chemisorbed on Fe/W(110) as a function of exposure. A reduction of ΔE_{ex} occurs with increasing dosage. After starting of the oxidation process (>12 L) the exchange splitting vanishes within the margin of error. Reprinted from [65], Copyright (1999), with permission from Elsevier

chemisorbed state into an oxygen-incorporated state; this incorporated two-dimensional thin oxide formed in the intermediate state is followed by the subsequent transformation into a bulk-like oxide with further oxygen exposure [68]. This explanation is corroborated by the observation that the exchange splitting decreases significantly between 6 and 10 L.

The observed exchange splittings [65, 69–71] originate from a spin dependent admixture of Fe $3d$ wave functions to the O $2p$ states. Theoretical calculations for oxygen on iron(100) demonstrate that the binding mechanism is mainly based on the hybridization of Fe $3d_{xy}$ and O $2p_{xy}$ states [66]. Additionally, the chemical bond between O $2p_z$ and Fe $3d_{3z^2-r^2}$ of iron atoms below the topmost layer plays a non-neglectable role. Exchange splittings of the order of 1 eV are predicted [66] for $p(1 \times 1) - Fe(100)$ as well as an induced magnetic moment on the oxygen atom with values of 0.24 μ_B [66] and 0.2 μ_B [72, 73]. Another theoretical calculation was carried out for $p(1 \times 1) - O/Fe(110)$ demonstrating a large induced magnetic moment of 0.7 μ_B as well as an exchange splitting of the O $2p$ derived band of about 1 eV [50, 74]. Although this (1×1) structure is not found on Fe(110) at room temperature, the observed exchange splittings are clearly indicating the presence of such a magnetic moment at the adsorbate atom. Additionally, after low exposures of oxygen the iron surface remains "magnetically alive" which is in agreement with these calculations.

Investigations of metallic adsorbates in the monolayer regime on ferromagnetic substrates also indicate the existence of magnetic moments within the overlayer (e.g. 3 μ_B for Mn/Fe(100) [75]) as well as differently strong hybridization for both spin channels [76].

Whilst the exchange split photoemission peaks suggest the presence of a magnetic moment on the adsorbate sites they do not allow any quantitative determination of the magnitude of that moment. This suggestion is in agreement with theoretical calculations which show that the hybridization can induce a sizeable magnetic moment on the adsorbed oxygen atom that results from the different occupancy of the adsorbate-induced majority and minority bands (see, e.g., [74]). Investigations of ferromagnetic materials, antiferromagnetic materials, spin glasses, and free atoms demonstrated that the exchange splitting is linearly correlated to the magnetic moment; a splitting of 1 eV results in a magnetic moment of 1 μ_B [77]. "Magnetically dead" layers at the surface are only created with the beginning of the incorporation of the oxygen atoms into the bulk and of the oxidation process, respectively. FeO is an antiferromagnetic insulator (a so-called Mott insulator) with the NaCl structure. All spins are oriented antiparallel between the nearest neighbors, i.e. a spin resolving photoemission experiment determines a vanishing spin polarization even below the Néel temperature. It should explicitly be noted that a magnetic moment within the oxygen overlayer is present for the *oxygen-on-iron system* exhibiting hybridization between adsorbate and substrate.

On most transition metal surfaces adsorption of oxygen leads to a broad structure in photoelectron spectra at a binding energy of about 6 eV caused by O $2p$ induced states. The system O/Co/W(110) additionally shows a peak at about 10 eV in normal emission (see Fig. 5.20). After subtraction of a linear

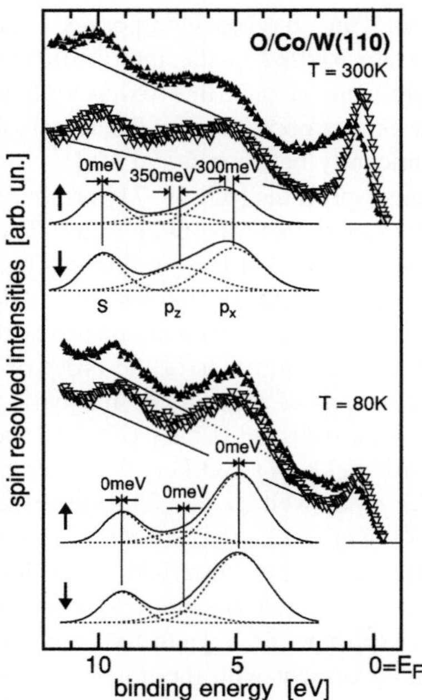

Fig. 5.20 Spin resolved photoelectron spectra of oxygen on Co/W(110) after an exposure of 20 L. The spectra were taken at $hv = 21.22$ eV and normal emission. Full symbols (*filled upward triangle*) denote the majority spin channel, open symbols (*open downward triangle*) the minority channel. The *upper spectra* give evidence of a coexistence of chemisorbed oxygen and antiferromagnetic CoO near the surface due to the satellite structure, induced by a charge transfer between O and Co, and the exchange splittings of the O $2p$ derived bands. After forming Co_3O_4 (*lower spectra*) no splittings are detectable. *Thin lines* represent a least-squares fit with Gaussian functions. Reprinted from [78], Copyright (1994), with permission from Elsevier

background, a least-squares fit with Gaussian functions was carried out to determine peak positions of adsorbate induced structures in the spin resolved spectra.

The results of this fit are represented in Fig. 5.20 by broken lines. Full symbols, (filled upward triangle) and $\uparrow$ denote the majority, open symbols (open downward triangle) and $\downarrow$ the minority spin channel. After an exposure of 20 L at different substrate temperatures it was found that the binding energies of the oxygen induced peaks are different, $E_B = 5.3, 7.3,$ and 9.9 eV at $T = 300$ K and $E_B = 4.9, 6.9$ and 9.2 eV at $T = 80$ K. On Co(0001) the oxygen molecules adsorb dissociatively. At room temperature O diffuses into the bulk and forms CoO [79–81]. Only very high doses lead to Co_3O_4 [79]. At temperatures below 120 K, however, the formation of cobalt oxide starts immediately with Co_3O_4 [80]. The binding energies measured in this experiment show good agreement with values for adsorbed oxygen on a recrystallized Co(0001) surface [80] (5.3, 6.7, and 10.1 eV at $T = 300$ K; 5.0, 6.6, and 9.3 eV at $T = 120$ K) and for cobalt oxide [82] (5.3, 7.3, and about 10 eV for CoO; about 5, 6.7, and about 10 eV for Co_3O_4) giving evidence for the presence of different oxidation states. The broad peak between 4 and 8 eV is caused by $O\ 2p$ derived bands, the other one at about 10 eV represents a satellite structure induced by a charge transfer from the occupied adsorbate state to unoccupied metal states with the correct symmetry ($O\ 2p \rightarrow Co\ 3d$) [83, 84].

The spin analysis demonstrates [78, 85, 86] that at $T = 300$ K for both states of the oxygen induced structure, formed by the p_z at about 7 eV and the p_x state at about 5 eV, exchange splittings of $\Delta E_{ex} = 0.35 \pm 0.1$ eV and $\Delta E_{ex} = 0.3 \pm 0.1$ eV are evident for p_z and p_x, respectively. These values are therefore larger than those for the $O\ 2p$ induced state of the system O/fcc–Co/Cu(100) with $\Delta E_{ex} = 0.2 \pm 0.1$ eV [87]. In contrast, the satellite structure at 10 eV binding energy reveals *no* splitting (upper panel of Fig. 5.20). Hence, at this coverage (exposure 20 L) the non-vanishing exchange splitting of the $O\ 2p$ induced states indicates, in connection with this satellite peak, a coexistence of chemisorbed oxygen and CoO near the surface. This behavior is explained by the observation that O diffuses into the bulk and forms antiferromagnetic CoO [80]. However, at the same time oxygen can be present in a chemisorbed state on the surface forming exchange split states by hybridization with cobalt bands. For all three features of the Co_3O_4 layer, prepared at 80 K, however, no peak splittings are visible (lower panel of Fig. 5.20). Nevertheless, a spin polarization was measured over the whole spectrum, indicating that parallel to the formation of Co_3O_4 also ferromagnetic cobalt is present at the surface.

The strong hybridization between adsorbate atom and ferromagnetic surface is a prerequisite for the induced magnetic moment in the adsorbate layer (as, e.g., additionally shown for iodine on Fe(110) [88]). More weakly bound species like CO and benzene do not exhibit an exchange splitting of their induced features [89, 90]. Surprisingly, the noble gas atom xenon "feels" the ferromagnetism of the underlying substrate [91, 92] due to the interaction of the ionic hole being created in the photoemission process.

As discussed in Sect. 5.1.2 it turned out that magnetic dichroism experiments are able to determine magnetic properties. In the following it will be shown that this experimental technique also allows the investigation of the magnetic behavior of adsorbate atoms on ferromagnetic surfaces [93].

In order to examine magnetic phenomena of adsorbates cobalt films being exposed to about 20 L of oxygen using MCDAD will now exemplarily be discussed. MCDAD experiments on oxygen are restricted to $O\ 2p$ levels since the O $1s$ and $O\ 2s$ levels exhibit no spin–orbit interaction (orbital momentum $\ell = 0$). The photoemission data displayed in Fig. 5.21 show the well-known oxygen induced $O\ 2p$ peak at binding energies around 6 eV. These features are characterized by the oxygen $2p_x$ on the low and the $2p_z$ orbital at the high binding energy side which exhibits a shift to a higher binding energy due to the chemical interaction with the Co $3d$ band. For a better visibility one spectrum (magnetization M^+) is always shifted upwards.

The full curves below the oxygen $2p$ orbital characterize the background, and the insets display the oxygen feature after background subtraction. For both systems the oxygen induced structures have been fitted with two Gaussian functions

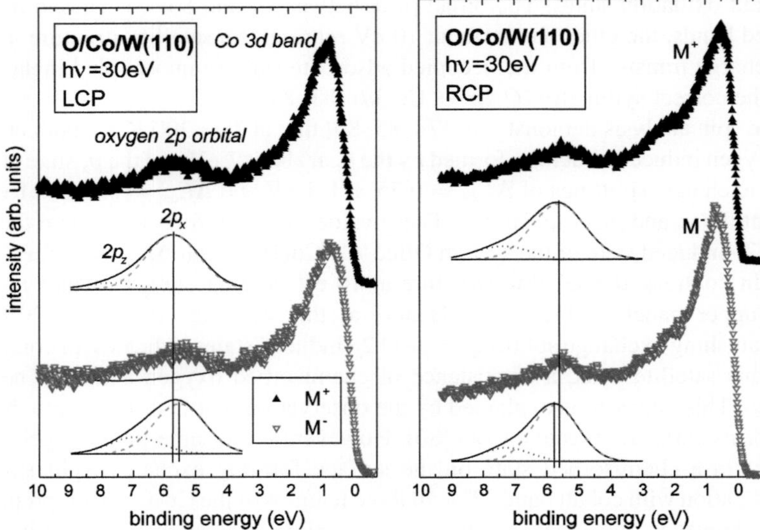

Fig. 5.21 *Left* normal emission photoelectron spectra ($hv = 30$ eV) from oxygen adsorbed on a thin hcp(0001) cobalt film on W(110) taken with left circularly polarized (*LCP*) light. Note that the data for the M⁺ magnetization direction are shifted upwards. The *insets* show the fit curves for both oxygen induced $2p$ orbitals (*dashed curves p_x, dotted curves p_z*) enlarged by a factor of two. The *solid lines* show the oxygen peak after background subtraction. *Right* corresponding photoemission spectra from oxygen on thin cobalt films but for right circularly polarized (*RCP*) radiation. Reprinted from [93], Copyright (1998), with permission from Elsevier

(dashed line p_x, dotted line p_z). The data clearly show a small splitting of about 200 meV with opposite sign for LCP and RCP radiation in the O $2p_x$ derived peaks depending on the magnetization direction. A splitting of the O $2p_z$ orbital can also be observed but the intensity is too small; thus giving a certain value of the exchange splitting makes no sense.

A detailed discussion of the symmetry properties of O/Co(0001) is difficult since oxygen adsorbs in a disordered state on this surface. In the case of an ordered structure (O/Fe(001)) Huang and Hermanson [66] calculated an induced magnetic moment of oxygen to be about 0.24 μ_B. Moreover, the photoelectron intensity differences (i.e. the MCDAD asymmetry) in the Co $3d$ band near the Fermi level decrease after oxygen exposure due to the chemical interaction of oxygen with the topmost cobalt layer. The reduced asymmetry displays a reduced magnetic moment of cobalt.

Taking the small value into account for the magnetic moment of oxygen from O/Fe(001) and additionally a value of less than 100 meV for the spin–orbit splitting it is astonishing that the MCDAD data really display a splitting of about 150 meV in the oxygen derived feature. However, it also clearly demonstrates the sensitivity of this technique. The presence of MCDAD phenomena in the oxygen derived bands shows that the adsorbed oxygen atoms indeed exhibit a non-vanishing spin–orbit and exchange interaction; otherwise the MCDAD effect would not appear.

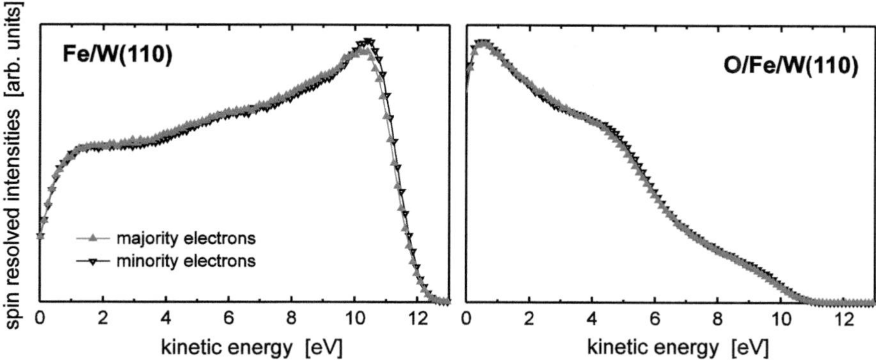

Fig. 5.22 Spin resolved MD spectra of clean and oxygen covered iron on W(110). The magnetization was carried out along the $[1\bar{1}0]$ direction

Panaccione et al. [94] applied the technique of magnetic linear dichroism (MLDAD) to the influence of chemisorbed adsorbate atoms (here, sulphur) on the magnetic moment carried in the topmost iron layer. Their MLDAD data from the Fe $3p$ level show a reduced asymmetry for the $c(2 \times 2) - S/Fe(001)$ structure originating from a smaller splitting in the m_j sublevels. Hence, MLDAD as well as MCDAD are useful methods for studying adsorbate induced magnetic properties.

To enhance the surface sensitivity SPMDS was used for the determination of the magnetic behavior of oxygen adsorbed on Fe(110) and Co(0001) [95, 96]. These results will be presented in the following.

The electron energy spectra of a clean Fe/W(110) film with a thickness of about 20 Å and of O on such Fe/W(110) films with an oxygen exposure of 3 L are shown in Fig. 5.22. The structure at high kinetic energies is caused by Fe $3d$ electrons near the Fermi level. After dosing oxygen to the iron surface, the emission of these electrons is drastically reduced.

A more detailed insight into the magnetic behavior is provided by the consideration of the corresponding asymmetry. Due to the high figure of merit of the $He(2^{3S})$ beam it is possible to determine the spin asymmetry with good statistical accuracy in energy steps of about 100 meV in order to look for weakly pronounced structures, in contrast to the pioneering studies with energy steps of 1 [97] and 2 eV [48], respectively. The asymmetry was measured in dependence of the oxygen coverage. These results are shown for iron in the left part of Fig. 5.23.

For the clean iron surface a *positive* asymmetry is measured for electrons near the Fermi level which corresponds to a dominance of the minority electrons in the surface-vacuum region. The asymmetry for electrons near the Fermi level (at a kinetic energy of about 11 eV) *changes its sign* after an oxygen exposure of about 3 L.

These observations are in agreement with measurements on O/Fe/GaAs(110) [48] and also with a theoretical calculation [74] which predicts a *high negative* spin density at E_F for the clean Fe(110) surface and a *positive* one after oxygen

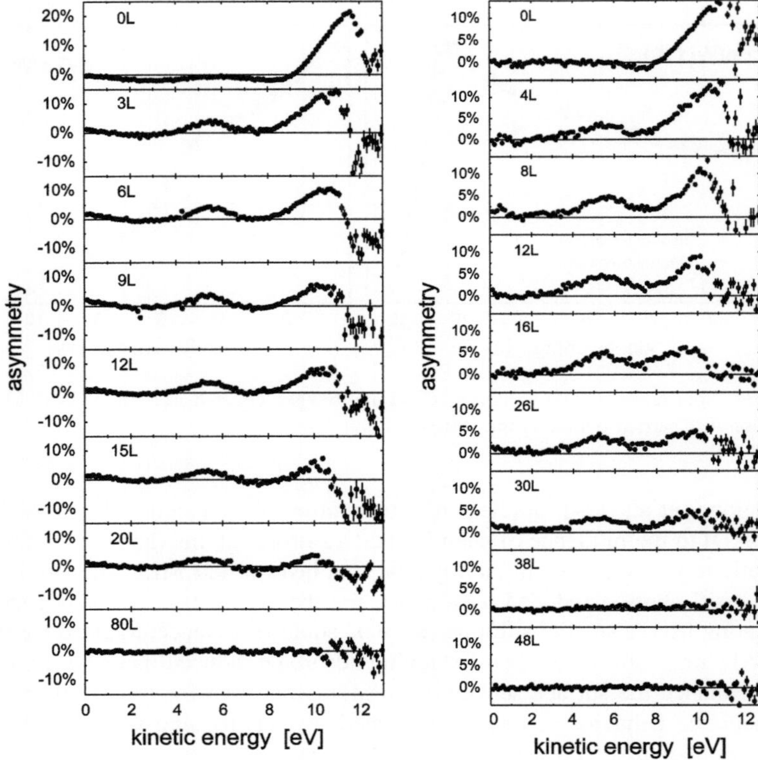

Fig. 5.23 *Left* spin asymmetry curve of Fe/W(110) for different oxygen coverages. At the Fermi edge a change in the sign of the asymmetry is present after an exposure of 3 L. *Right* spin asymmetry curve of Co/W(110) for increasing oxygen coverages. A change in the sign of the asymmetry at the Fermi edge does *not* occur being in contrast to the oxygen exposed iron surface

adsorption (see Fig. 5.24). It should be noted that this calculation is based on a p(1 × 1) structure of the oxygen overlayer, whereas experimentally c(2 × 2) and c(3 × 1) structures are present with increasing exposure [68].

Up to an exposure of about 3 L, the iron related asymmetry decreases monotonously from its maximum and the oxygen related asymmetry increases to a maximum. Between 3 and 12 L, a constant negative asymmetry of iron and a rapidly to zero decreasing one for the oxygen related structure near 5.5 eV occurs. This observation is interpreted such that chemisorbed oxygen has been transformed to FeO as already discussed above.

The spin resolved de-excitation spectra of oxygen on Co(0001) surfaces exhibit a similar behavior to O/Fe(110) with a distinct decrease in intensity near the Fermi level for increasing oxygen exposure (not shown here; see [95, 96]). The spin asymmetry in dependence of oxygen exposure to Co/W(110) is presented in the right part of Fig. 5.23. In the oxygen induced structure at about 5.5 eV, the asymmetry increases from zero and remains constant at about 5% above 8 L.

Fig. 5.24 Layer projected density of states for oxygen on Fe(110) (*top*) and clean Fe(110) (*bottom*) (reprinted with permission from [74]. Copyright 1993, American Institute of Physics). The *solid lines* represent majority and *dashed lines* minority electrons. Fe(S) denoted the topmost Fe layer, Fe(C) the iron center layer, and O the oxygen adsorbate overlayer

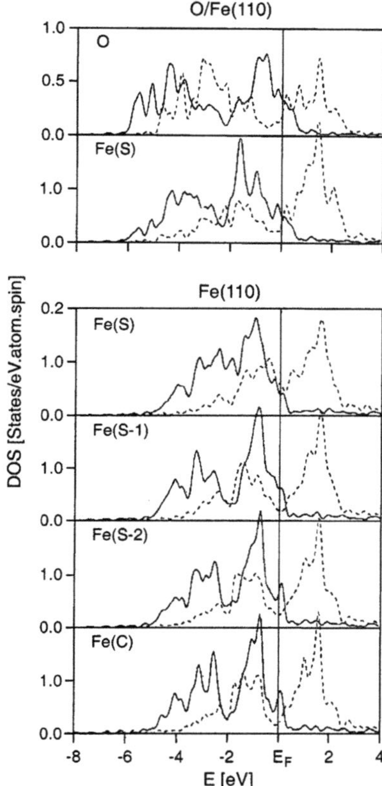

In the cobalt induced structure (at a kinetic energy of about 11 eV) the asymmetry decreases *without changing sign* with increasing exposure. This means that minority electrons dominate directly at the surface for both the bare and the oxygen covered one. Therefore, this behavior is markedly different from iron.

5.2 Ferromagnetic Materials with Localized Magnetic Moments

In contrast to the commonly known itinerant ferromagnetic materials Fe, Co, and Ni, the valence electrons carry the magnetic moment *and* are simultaneously responsible for the exchange coupling, the rare earth metals are so-called local moment systems, i.e. the magnetic moment is localized in the atomic $4f$ shell. Gadolinium (Gd) with the electron configuration $[Xe]\ 4f^7\ 5d^1\ 6s^2$ has often been regarded to as a prototype rare earth metal since it exhibits a half filled $4f$ shell which carries— according to Hund's rules—the maximum possible f shell magnetic moment of 7 μ_B. Below the Curie temperature of 292.5 K Gd is ferromagnetic. The atomic $4f$ moments are exchange coupled via the Ruderman–Kittel–Kasuya–Yosida (RKKY) interaction mediated by the $5d$-like conduction electrons which also give a small contribution to the total magnetic moment of about 0.6 μ_B. However, the detailed relationship between the electronic structure and magnetism of the bulk and—even more drastically—the surface have controversially been discussed. The discussion was triggered off by a combined MOKE and spin polarized low-energy electron diffraction (SPLEED) study on the magnetism of the Gd(0001) surface by Weller et al. [98]. The authors claimed to observe an enhanced surface Curie temperature T_{CS} and a non-ferromagnetic, possibly antiferromagnetic, coupling between the bulk and the surface layer.[1]

These extraordinary surface magnetic properties remained unexplained until Freeman et al. [99] proposed the existence of a highly localized d_{z^2}-like surface state. According to this calculation the surface state is exchange split into an occupied majority and an empty minority branch. This difference in occupation was made responsible for an enhanced surface magnetic moment and T_{CS}. Indeed, the existence of the occupied majority part of the d_{z^2}-like surface state was soon experimentally confirmed by angle resolved photoemission spectroscopy [100]. Later on also the empty minority part of the surface state was found by inverse photoemission (IPE) [101]. Further predictions based on this calculation, however, turned out to be incorrect, namely the expansion of the outermost interlayer spacing of Gd(0001) by $\approx 6\%$ and an antiferromagnetic coupling between bulk and surface. Instead, a quantitative LEED study showed that the outermost interlayer spacing is contracted by $3.5 \pm 1.0\%$ [102].

Furthermore, it is nowadays opinion that bulk and surface of Gd(0001) couple ferromagnetically instead of antiferromagnetically [103–105]. As described in detail by Bylander and Kleinman [106] the discrepancy between theoretical and experimental data might be caused by a general underestimation of exchange, correlation, and magnetic energy within the FLAPW method which cannot be corrected in the case of Gd.

5.2.1 Temperature Dependent Characteristics

5.2.1.1 Temperature-Dependent Electronic Properties of Gd(0001)

In the past the effort to improve the understanding of finite temperature magnetic properties has mainly concentrated on itinerant ferromagnetic materials like Fe, Co, and Ni. The theoretical starting point was the one-electron finite temperature band theory which is known under the name "Stoner-theory" [108, 109]. Within

---

[1] Since single crystalline surfaces of rare earth metals are extremely difficult to clean almost all experiments have been performed on thin Gd films with the epitaxial relationship $(0001)Gd\|(110)W$ and $[11\bar{2}0]Gd\| [1\bar{1}0]W$ [107]. It was found that Gd(0001)/W(110) is fully relaxed at a coverage of approximately 35 monolayers (1 ML $\simeq$ 2.89 Å).

this theory, the characteristic parameter is the exchange splitting ΔE_{ex}, defined as the energetic difference between majority and minority spin bands. According to this theory ΔE_{ex} decreases with increasing temperature until majority and minority spin bands merge at the Curie temperature T_{C}. Simultaneously, the magnetic moments disappear. Stoner's theoretical approach was dedicated to quasi-free electrons, a condition which is obviously not fulfilled by the narrow energy d bands that are responsible for magnetism in the itinerant magnets. Instead, strong correlation effects have to be considered which are included in the framework of the "local-band theory" by local moments that remain at an almost constant amplitude but exhibit transverse fluctuations [110–112].

In the framework of the local-band theory short-range spin order may persist even above T_{C} although *per definitionem* long-range spin order is lost at the ferromagnetic–paramagnetic phase transition. Indeed, transverse spin fluctuations are known to be an important mechanism in ferromagnetic transition metals and have been observed on Ni(110) [113]. Nowadays, it is widely accepted that the question whether or not the exchange splitting collapses in itinerant ferromagnetic materials at or above T_{C} substantially depends on the degree of localization of the considered electron bands which has to be compared with the size of regions that exhibit short-range spin order. As a consequence, ΔE_{ex} may exhibit a striking **k** dependence [114, 115].

In contrast for magnetic materials with localized magnetic moments, only very few data are available. Experimental effort to obtain a reliable data base of the temperature-dependent exchange splitting has concentrated on the (0001) surfaces of the rare earth (RE) metals Gd [103, 116–121] and Tb [122]. While Gd is ferromagnetic below its bulk Curie temperature $T_{\mathrm{CB}} = 293$ K, the bulk material of Tb exhibits three magnetic phases: (1) it is ferromagnetic below $T_{\mathrm{CB}} = 220$ K, (2) antiferromagnetic with a helical magnetic structure between T_{CB} and the bulk Néel temperature $T_{\mathrm{CN}} = 228$ K, and (3) paramagnetic above T_{CN}. In the case of Gd(0001) it has consistently been described by experimentalists that the bulk majority and minority spin bands, occupied [118, 121] as well as empty ones [103, 117], merge at the bulk Curie temperature $T_{\mathrm{CB}} = 293$ K. This Stoner-like behavior might be caused by their itinerant character.

However, contradictory experimental results have been published concerning the exchange splitting of the RE(0001) surface state. This surface state exhibits a d_{z^2}-like symmetry, is strongly localized and was made responsible for so-called extraordinary surface magnetic properties [99, 123, 124]. While it was claimed for Gd(0001) in a spin polarized photoelectron spectroscopy (PES) study that the occupied majority part of the surface state exhibits no energetic shift with increasing temperature but instead looses spin polarization (so-called spin mixing behavior) [121], another combined inverse photoelectron spectroscopy (IPES) and PES study came to the result that occupied majority and empty minority spin bands merge together [116]. Surprisingly enough, the same group applying the same experimental techniques reported on spin mixing behavior for the surface state of Tb(0001) [122], a material which exhibits electronic properties being very similar to Gd(0001) [125, 126].

In the following the focus will be on the temperature dependence of the surface electronic structure of Gd(0001) and Tb(0001) as studied by scanning tunneling microscopy (STM) and scanning tunneling spectroscopy (STS). This dependence is discussed in detail for different morphologies of Gd samples and subsequently compared between Gd and Tb.

Experiments performed by means of photoemission (PE) and inverse photo-emission (IPE) spectroscopy suffer from the limitation to electronic states below or above the Fermi level, respectively, i.e. the exchange splitting[2] $\Delta E_{\mathrm{ex}}^{\mathrm{surf}}$ can only be determined by using different experimental techniques not being carried out on the *same* sample. This experiment-inherent disadvantage carries additional weight as this, first, multiplies potential systematic errors which may occur in the determi-nation of the Fermi edge in the spectra and, second, as the complex film morphology of Gd may cause different magnetic properties. Scanning tunneling spectroscopy, in contrast, avoids both problems since it enables to detect the density of electronic states above sample surfaces below the tip apex on *both* sides of the Fermi level in a *single* measurement and *spatially resolved*. Since the $5d_{z^2}$-like surface state is located around the $\bar{\Gamma}$–point of the surface Brillouin zone it exhibits a large fade-out length into the vacuum and is therefore ideally suited to be investigated by STS. Therefore, the temperature dependence of the exchange splitting of the Gd surface state can *directly* be determined. Additionally, the spatial resolution enables to evaluate the influence of morphology and local film thickness.

As mentioned above it is still under debate whether or not the binding energy of both parts of the surface state changes with temperature. To unravel this open question scanning tunneling spectroscopy was performed at variable temperatures [119, 120, 128–134]. Figure 5.25 shows typical tunneling spectra as measured above the monolayer (gray curve) and three-dimensional islands (black curve) of a Gd film grown in the Stranski–Krastanov mode at four selected temperatures: close to the ground state ($T = 29$ K), slightly cooled ($T = 171$ K), at room temperature ($T = 293$ K), and at elevated temperature ($T = 357$ K). Close to the ground state ($T = 29$ K), both peaks being characteristic for the occupied and empty part of the surface state can clearly be observed at $U = 460$ mV and $U = -240$ mV, respectively.

As the temperature is increased from $T = 29$ K up to 293 K both peaks obviously shift towards the Fermi level, i.e. zero bias. This observation is in strong disagreement with a pure spin mixing behavior as proposed on the basis of PE experiments performed by Li et al. [121]. However, increasing the temperature above 293 K does not lead to a further shift of both peaks. Unfortunately, the binding energy of the occupied surface state could not be determined above $T = 360$ K. This is caused by the background of the differential conductance

---

[2] Although it has been questioned by Sandratskii and Kübler [127] whether the energetic difference between the peak positions of the occupied and the unoccupied surface state can be considered as an exchange splitting $\Delta E_{\mathrm{ex}}^{\mathrm{surf}}$; in the following this terminology will be used.

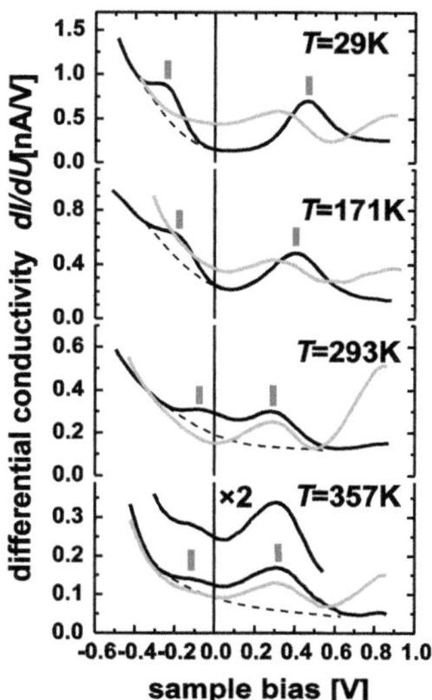

Fig. 5.25 Typical tunneling spectra measured above Gd(0001) islands (*black*) and 1 ML Gd/W(110) (*gray*) at 29, 171, 293, and 357 K (topography similar to Fig. 3.6b). Close to the ground state ($T = 29$ K) the exchange splitting between the occupied and unoccupied surface state above Gd(0001) islands is ≈ 700 meV. While the binding energy of the electronic feature measured above the Gd monolayer remains unchanged, the exchange splitting of the surface states decreases with increasing temperature. Reprinted from [129], Copyright (1999), with permission from Elsevier

which rises to negative sample bias as indicated by a hatched line in Fig. 5.25. As the temperature is increased this background extends its tail above the Fermi level. Thereby, it becomes more and more difficult to determine the peak position of the occupied surface state and the shoulder vanishes above 360 K. For comparison the monolayer spectra are plotted, too. The corresponding peak does not shift significantly in the temperature range under study.

Figure 5.26 summarizes the temperature dependent results on the binding energy of the occupied and the unoccupied part of the surface state, respectively, as measured above Gd films of different local coverage Θ: Gd films with $\Theta > 30$ ML (asterisk), Gd islands with 4 ML $< \Theta < 15$ ML (filled square), bilayer islands (open circles), and the Gd monolayer (open diamond). In the temperature range between 20 and 380 K the binding energy of the unoccupied (occupied) part of the surface state as measured above Gd islands filled square shifts from 490 meV (-220 meV) towards 300 meV (-60 meV). In contrast, the peak position of the unoccupied electronic state above the first monolayer remains unchanged (cf. Fig. 5.25). These results are in good agreement with an earlier temperature dependent PE and IPE study [116] up to 300 K

Within energetic resolution ($\pm$ 50 meV) the high temperature STS data also fit PE measurements from the same reference. However, a significant discrepancy exists for IPE results obtained above 300 K [116, 117]. The STS data reveal that

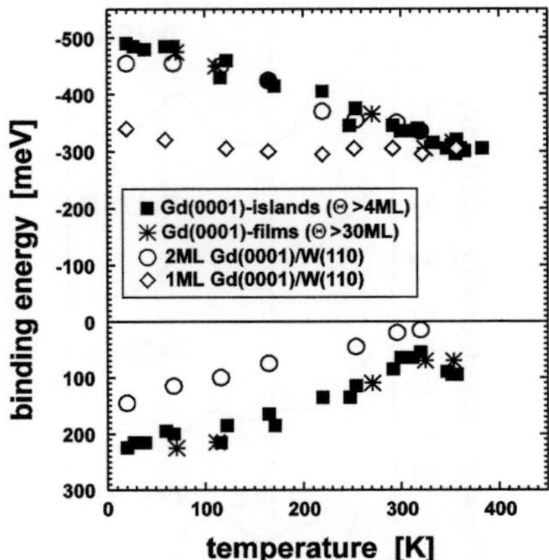

Fig. 5.26 Plot of the binding energy versus sample temperature for surface states as measured above (*filled square*) thick Gd islands ($\Theta \geq 5$ ML), (*asterisk*) Gd films ($\Theta \geq 30$ ML), (*open circle*) Gd bilayers, and (*open diamond*) the Gd monolayer. Every data point represents an average of several measurements. The overall error bars are estimated to be ± 50 meV. The binding energy of the electronic feature above the Gd monolayer remains constant within the experimental error. The exchange splitting between occupied and unoccupied surface states is smaller for the bilayer than for thicker films. This is mainly caused by a reduced binding energy of the occupied surface state. Reprinted from [119], Copyright (1998), with permission from Elsevier

the binding energy of the unoccupied surface state is 300 meV instead of 20 meV above 350 K [116]. The results indicate no difference between *dI/dU* spectra of smooth Gd films (asterisk) and Gd islands (filled square). Furthermore, samples as shown in Fig. 3.6c enabled to measure also the peak positions above Gd bilayer islands on W(110) (open circle). While the binding energy of the unoccupied surface state is basically the same for bilayer islands and three-dimensional Gd islands, the binding energy of the occupied surface state is diminished by approximately 100 meV at temperatures below 250 K (cf. Fig. 3.12) and by approximately 50 meV above room temperature.

The temperature dependence of the exchange splitting $\Delta E_{\text{ex}}^{\text{surf}}$ between the peak positions of the occupied and the unoccupied surface state, as measured above various Gd films (cf. Fig. 3.6a–c), is plotted in Fig. 5.27. For thick Gd films the splitting amounts to about 700 meV close to the ground state. The splitting reduces with increasing temperature in the temperature range between 20 and 300 K but does *not* approach zero up to the maximum temperature in this study, i.e. 360 K. The exchange splitting measured above the two atomic layers high Gd islands amounts to only 600 meV at 20 K and also reduces continuously down to approximately 370 meV at 300 K. While the transition from decreasing to constant exchange splitting might be interpreted in terms of a Curie temperature T_{C} this is strictly excluded for a film thickness as low as 2 ML since at this coverage T_{C} is far below 100 K [135].

Obviously, the experimental data can neither be described within a pure Stoner nor within the pure spin mixing picture. However, one should keep in mind that

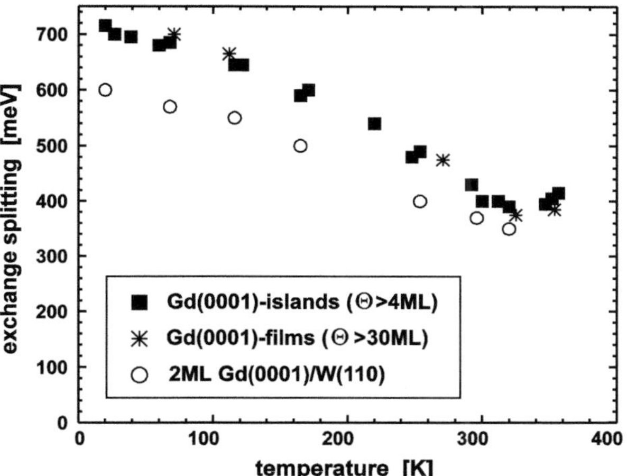

Fig. 5.27 Temperature dependent exchange splitting of the Gd(0001) surface state. Reprinted from [119], Copyright (1998), with permission from Elsevier

the magnetism of Gd is caused by localized magnetic moments resulting from the half-filled $4f$ shells which are coupled via RKKY interaction mediated by spin polarized $5d$ conduction electrons. At least qualitatively the differences in the temperature dependent behavior of the exchange splitting between $5d$ bulk and surface states can be explained by their different degree of localization [121]. While the bulk $5d$ conduction electrons are delocalized the $5d_{z^2}$-like surface state is highly localized even within the surface plane [99, 101]. As already described by Sandratskii and Kübler [127] the magnitude of the atomic $4f$ magnetic moment is constant with temperature. In the ground state Gd is ferromagnetic and all $4f$ moments are aligned parallel. Rising the temperature leads to fluctuations of the relative axes of adjacent moments which have been theoretically modeled in the past by spin waves with the wave vector $\mathbf{q}$, i.e. the atomic moment of the $(n + 1)$th atom is rotated by a certain angle α with respect to the nth moment (see Fig. 5.28). In this situation the degree of localization is of importance: the less its localization the more the $5d$ electron experiences the collective $4f$ polarization which decreases the resulting exchange splitting until it vanishes when the so-called magnetic correlation length exceeds the spiral periodicity $|\mathbf{q}|^{-1}$.

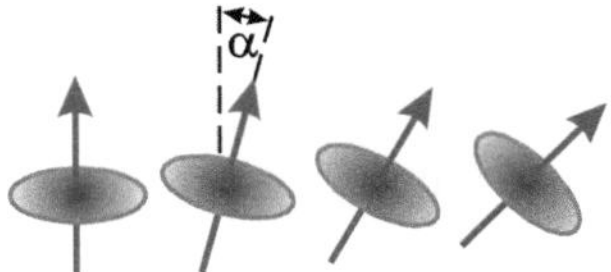

Fig. 5.28 The $4f$ shell of each Gd atom carries a local magnetic moment which remains constant with temperature. The relative axes of these magnetic moments fluctuate with rising temperature. This behavior can be modeled by spin waves being characterized by the angle α between nearest neighbors. From [127], copyright 1998, reproduced with permission from World Scientific Publishing Co. Pte. Ltd

The experiments show that the exchange splitting of the Gd(0001) surface state is reduced with increasing temperature between 20 K and approximately 300 K but does not vanish up to a temperature of 360 K being well above any possible surface Curie temperature. In the framework of the spin wave model already mentioned above and considering only nearest neighbor interactions—which is sufficient in most cases—this observation suggests that the $4f$ moments of adjacent Gd atoms turn their relative quantization axis from parallel ($\alpha = 0$) to an unordered configuration ($\langle\alpha\rangle = \pi/2$) between 20 and 300 K thereby going from minimum to maximum disorder. Increasing the temperature further has no effect on the spin order between nearest neighbors and consequently the exchange splitting remains constant since the high spatial localization of the surface state helps to maintain a local exchange splitting due to interaction with the atomic $4f$ magnetic moments at temperatures which exceed the Curie temperature. The same argument can also be applied for the results obtained on Gd films as thin as 2 ML.

In the following the results for Gd(0001) with a thickness of more than four layers will be compared to Tb(0001) which additionally possesses an antiferro-magnetic state at temperatures between 220 and 228 K.

By increasing the sample temperature both peaks of the Tb(0001) surface state shift towards E_F thereby decreasing the exchange splitting ΔE_{ex}. Just above T_{CB} ($T = 223$ K) the occupied part of the surface state has already approached the Fermi level very closely [$U = -50 \pm 34$ mV]. In contrast to earlier experiments on Gd(0001), this trend continues for Tb(0001) even above its bulk magnetic phase transition temperatures $T_{CB} = 220$ K and $T_{NB} = 228$ K. At $T = 248$ K the former occupied part of the surface state is energetically localized at the Fermi level [$U = 10 \pm 30$ mV]. Increasing the temperature further ($T = 258$ K and $T = 271$ K) the maximum in the dI/dU spectra crosses E_F. As a result the surface state which was clearly occupied at low temperature ($T = 85$ K) becomes partially empty above $T = 250$ K.

In Fig. 5.29 the temperature dependent peak positions of the "filled" and empty part of the Tb(0001) surface state are summarized. Furthermore for comparison, the data obtained on Gd(0001) have been included. Different temperature scales for Gd (top axis) and Tb (bottom axis) were used such that the data can be compared in terms of a reduced temperature T/T_{CB}. The temperature dependent behavior of Tb(0001) can be described by separating three different temperature regimes. In the first regime between $T = 16$ K and T_{CB} both, the empty as well as the occupied part of the surface state, shift only slightly, by about 100 meV, towards E_F. Second, increasing the temperature above T_{CB} leads to a more rapid shift towards E_F. Within the narrow temperature range $T_{CB} < T < 265 \pm 5$ K both parts of the surface state exhibit a further shift of about 120 meV. This leads to the fact that the maximum of the occupied part of the surface state crosses E_F at about 250 K, thereby becoming partially unoccupied. In a third regime above about 265 K the peak positions remain constant within the error bar.

When the temperature dependent surface electronic behavior of Tb(0001) is compared to that of Gd(0001), the most apparent qualitative difference is that the second temperature regime—being characterized by a rapid shift of both parts of

Fig. 5.29 Plot of the peak positions of the Tb(0001) surface state versus the sample temperature (*bottom axis*). While the empty part (*filled square*) of the surface state shifts from +440 mV at 85 K down to +180 mV at 360 K the occupied part (*filled circle*) moves from −150 mV at 85 K to about +40 mV at 270 K thereby crossing the Fermi level at about 250 K. For comparison data obtained on Gd(0001) are included (*top axis*). Reprinted with permission from [133]. Copyright (1999) by the American Physical Society

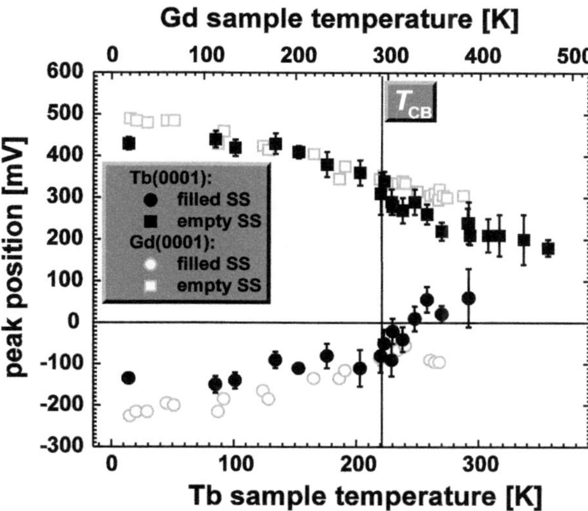

the surface state towards E_F above T_{CB}—does not exist for Gd(0001). Instead, the first temperature regime is directly followed by the third, i.e. by increasing the Gd(0001) sample temperature above T_{CB} peak positions are found that remain constant within the measurement accuracy. The difference becomes even more obvious when comparing the surface state exchange splitting ΔE_{ex} of Gd(0001) and Tb(0001) as shown in Fig. 5.30. Close to the ground state Tb exhibits a ΔE_{ex} of 600 meV while it amounts to 700 meV for Gd(0001), thereby reflecting their ratio of the local $4f$ moment of 7 μ_B in the case of Gd and 6 μ_B in the case of Tb. ΔE_{ex} decreases with increasing temperature down to about 400 meV at T_{CB} for both, Gd and Tb. However, increasing T above T_{CB} leads to a further reduction of ΔE_{ex} for Tb(0001), while it remains constant for Gd(0001). Only for $T > 260$ K the data obtained on Tb(0001) suggest a constant value of ΔE_{ex}.

This striking difference between the temperature dependent surface electronic properties of Tb(0001) and Gd(0001) is strongly related to the question which type of short-range spin order persists above T_{NB} and T_{CB}, respectively, i.e. helical antiferromagnetic (AFM) or ferromagnetic (FM) order. In this context it is important that the d_{z^2}-like surface state that exists on rare earth(0001) surfaces is strongly localized at one Gd surface atom with a small overlap to nearest neighbors (NN) underneath [99]. Therefore, the amount of exchange splitting is dominated by the interaction with the $4f$ moments of this particular atom and its NN. Below T_{CB} and T_{NB} the $4f$ moments exhibit parallel and helical order, respectively, as schematically presented in column I and II of Fig. 5.30b. It is well-known that many RE metals exhibit short-range spin order in the paramagnetic (PM) state. In the case of bulk Tb—as well as in other RE-metals like Dy and Ho—"AFM order persists $\approx$ 40 K above T_{NB}" [136]. This value almost perfectly fits to the temperature which a constant exchange-splitting ΔE_{ex} is found above being 260 K (see Fig. 5.30a). While the helical turn angle ω of Tb amounts to 20°

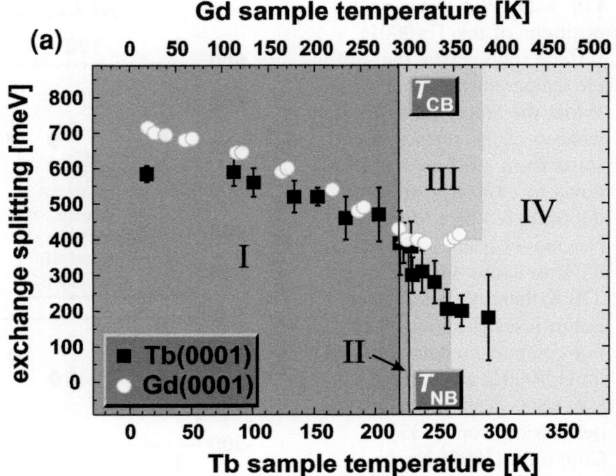

(b)	I	II	III	IV
	FM	AFM ($\alpha=20°$)	SR-AFM ($\alpha<40°$)	PM
Tb				
Gd		—		
			SR-FM	

Fig. 5.30 **a** Temperature dependent exchange splitting of the Tb(0001) (*filled square*) and Gd(0001) (*open circle*) surface state. **b** Schematic spin-structures of Tb and Gd: *I* ferromagnetic (*FM*), *II* antiferromagnetic (*AFM*), *III* short-range AFM and short-range FM, and *IV* paramagnetic (*PM*). Reprinted with permission from [133]. Copyright (1999) by the American Physical Society

between adjacent basal planes within the AFM phase it raises up to 40° between T_{NB} and 260 K, thereby stabilizing short-range AFM by ω being commensurate with the hexagonal lattice of Tb [136].

In contrast, Gd is the only magnetic RE metal that exhibits no spiral magnetic phase. Instead, strong evidence exists that short-range FM order persists "up to at least 340 K", possibly up to 400 K [137]. A spin polarized photoelectron study revealed that the Gd(0001) surface exhibits short-range (>20 Å) FM spin order up to about 380 K, too [138]. Keeping this in mind it is straightforward to explain the striking difference between Tb(0001) and Gd(0001) in regime III of Fig. 5.30a: the exchange splitting ΔE_{ex} of the Gd(0001) surface state remains constant above T_{CB} up to at least 360 K—probably up to 380 K—because it remains ferromagnetically ordered on the relevant length scale, i.e. several lattice constants. In contrast, the Tb short-range spin order between NN atoms decreases because ω increases from 20° at 228 K up to 40° at 260 K. Finally, in the PM regime (IV) any spin order is lost. Nevertheless, due to the high spatial localization of the surface state which still interacts with the atomic $4f$ magnetic moments a small local ΔE_{ex} is maintained.

It remains, however, the question whether Tb(0001) exhibits an enhanced T_{CB} as claimed by Rau et al. [124] on the basis of electron capture spectroscopy (ECS). The results of this study show that the electron spin polarization remains non-zero up to 248 K, but shows a striking non-monotonic behavior, i.e. it increases with increasing temperature above 240 K and peaks at $\approx$ 245 K. In this context it is worthwhile, however, to mention that ECS favors electronic states that are energetically located close to E_F with the momentum vector k roughly along the surface normal [139]. Furthermore, the ECS experiments have not been performed in remanence but in an external magnetic field between 25 and 600 Oe. Therefore, regions that exhibit short-range order are forced into a parallel alignment above T_{CB}. At about $T = 250$ K the "occupied" part of the Tb(0001) surface state which is localized around the $\bar{\varGamma}$ –point of the surface Brillouin zone crosses E_F (cf. Fig. 5.29) thereby fulfilling both conditions for an effective measurement by ECS which results in a peak of the spin polarization.

5.2.2 Magnetic Domain Imaging on the Nanometer Scale

The investigation of the exchange split surface state being discussed above was carried out with non-magnetic probe tips in the STM. In the following the extension to obtain magnetic information on the nanometer scale will be discussed by replacing the non-magnetic by ferromagnetic tips [131, 140–142].

A comprehensive overview and detailed discussion concerning spin polarized scanning tunneling microscopy is given in [143].

In the past many attempts have been made to obtain magnetic information by making the STM sensitive to the spin of the tunneling electrons. Basically two different concepts have been used to achieve spin polarized vacuum tunneling:

- By using magnetic STM probe tips the spin valve effect [144] can be exploited which relies on the fact that the tunneling conductance between two ferromagnetic electrodes separated by an insulating barrier depends on whether the magnetic moments are directed parallel or antiparallel. This effect has extensively been studied in planar tunneling junctions [145–148] and has been used in STM to probe the topological antiferromagnetic order of a Cr(001) surface by means of a CrO_2 tip [149].

  Another approach was carried out by periodically switching the magnetization direction of an amorphous CoFeNiSiB tip with a small coil wound around the tip [150–156] thus being sensitive to the perpendicular component of the sample magnetization. In order to realize an in-plane sensitivity the same type of technique is applicable. But, the "tip" now consists of a ferromagnetic ring whose magnetization is also periodically switched by a coil [157, 158].

- Optically pumped GaAs tips or samples enable spin polarized vacuum tunneling to be observed [159, 160]. However, magnetic domain imaging has not unambiguously been demonstrated yet.

All these experiments are limited by the need to separate topographic, electronic, and magnetic information in the case of magnetic probe tips, and to eliminate thermal or film thickness induced effects in the case of semiconducting tips.

The experimental approach to overcome these difficulties is based on tunneling into the surface state of Gd(0001) which is exchange split into a filled majority and an empty minority spin contribution [103, 121]. In analogy to the low-temperature experiments performed with ferromagnet–insulator–superconductor planar tunneling junctions [161, 162] where the quasiparticle density of states of superconducting aluminum is split by a magnetic field into spin up and spin down parts, two spin polarized electronic states with opposite polarization were used to probe the magnetic orientation of the sample relative to the tip [139].

5.2.2.1 Spin Polarized Tunneling into Exchange Split Surface States

Figure 5.31 schematically illustrates the principle of spin polarized scanning tunneling spectroscopy (SPSTS). A sample which exhibits an exchange split surface state with a relatively small exchange splitting is ideally suited for the experimental approach. If the exchange splitting ΔE_{ex} is too large one spin component would be too far from the Fermi level and not accessible by STS, as e.g. in the case of Fe(001), where ΔE_{ex} amounts to 2.1 eV and only the minority band appears as a peak in the dI/dU spectra just above the Fermi level [163]. In contrast, the majority (minority) part of the Gd(0001) surface state at 20 K has a binding energy of -220 meV (500 meV), i.e. the exchange splitting only amounts to 700 meV far below the Curie temperature of 293 K.

In the following, vacuum tunneling between a Gd(0001) surface and a tip material is considered which the sign of the spin polarization does not reverse for in the energy range of interest, i.e. ± 0.5 eV around the Fermi level. This condition is fulfilled for Fe [50]. For simplicity, a constant spin polarization is assumed (see Fig. 5.31a, bottom). If the magnetization direction of the tip remains constant two possible magnetic orientational relationships between tip and sample occur, parallel or antiparallel. Since, however, both the majority and the minority component of the Gd(0001) surface state appear in the tunneling spectra, in any case the spins of *one* component of the surface state will be *parallel* with the tip while the *other one* will be *antiparallel*. Therefore, the spin valve effect will differently act on the two spin components; due to the strong spin dependence of the density of states the spin component of the surface state parallel to the tip magnetization is enhanced at the expense of its counterpart being antiparallel. Consequently, by comparing tunneling dI/dU spectra measured above domains with opposite magnetization one expects a reversal in the contrast at the majority and minority peak position (see Fig. 5.31b).

Tunneling spectra measured in an external magnetic field with an Fe coated probe tip ($\Theta_{\mathrm{tip}} = 10$ ML) positioned above an isolated Gd(0001) island show exactly the expected behavior (see Fig. 5.31c). After inserting the sample in the STM sample holder and cooling down to 70 K it was magnetized in a magnetic field of $+4.3$ mT applied parallel to the sample surface. Subsequently, 128 tunneling dI/dU spectra were measured in remanence with the tip positioned above the Gd island marked by an arrow in the inset of Fig. 5.31c. Then the direction of

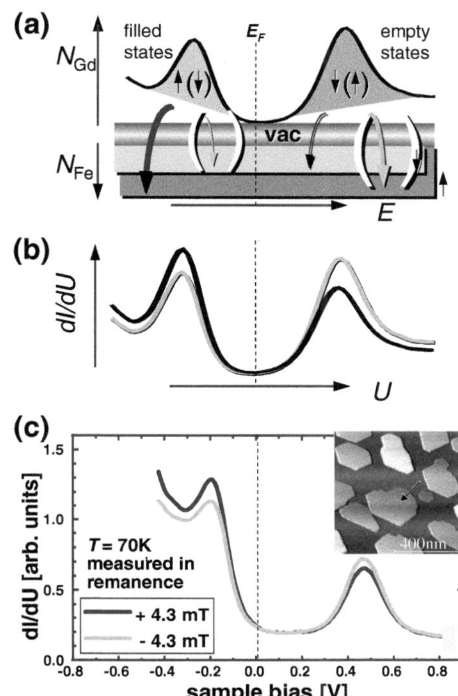

Fig. 5.31 a The principle of SPSTS using a sample with an exchange split surface state, e.g. of Gd(0001), and a magnetic Fe tip with a constant spin polarization close to E_F: due to the spin valve effect the tunneling current of the surface state spin component being parallel to the tip is enhanced at the expense of its spin counterpart. **b** This should lead to a reversal in the *dI/dU* signal at the surface state peak position upon switching the sample magnetically. **c** Exactly this behavior could be observed in the tunneling spectra measured with the tip positioned above an isolated Gd island (see *arrow* in the *inset*). Reprinted with permission from [163]. Copyright (1998) by the American Physical Society

the magnetic field was reversed (−4.3 mT) and further 128 tunneling *dI/dU* spectra were measured at the same location. This procedure was repeated several times. The schematic experimental setup is given in Fig. 5.32. Figure 5.31c shows the averaged tunneling spectra measured in remanence after the application of a positive or negative field. Comparison of the spectra reveals that for positive field the differential conductance *dI/dU* measured at a sample bias which corresponds to the binding energy of the occupied (majority) part of the surface state is higher

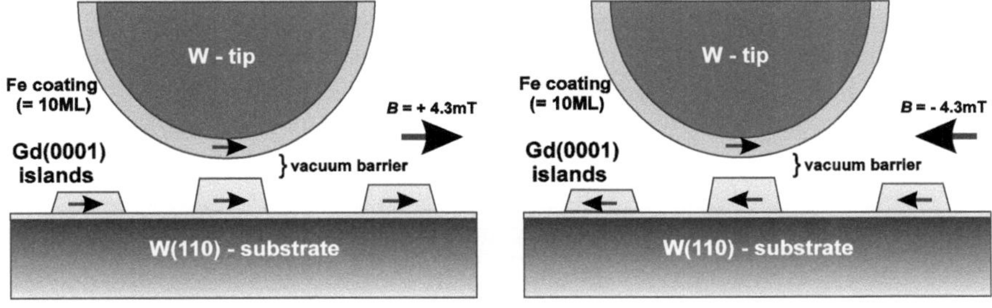

Fig. 5.32 Schematic drawing of the experimental setup for measuring spin polarized vacuum tunneling between an Fe coated tip and free standing Gd islands (reprinted with permission from [132]. Copyright 1999, American Institute of Physics)

Fig. 5.33 *dI/dU* image at the **a** majority ($U = -0.2$ V) and the **b** minority ($U = +0.45$ V) surface state peak position. While a strong contrast within the Gd island can be recognized in **a**, a weaker and opposite contrast is present in **b**. **c** In the asymmetry image the contrast is enhanced and tip changes less visible. Reprinted with permission from [130]. Copyright (1998) by the American Physical Society

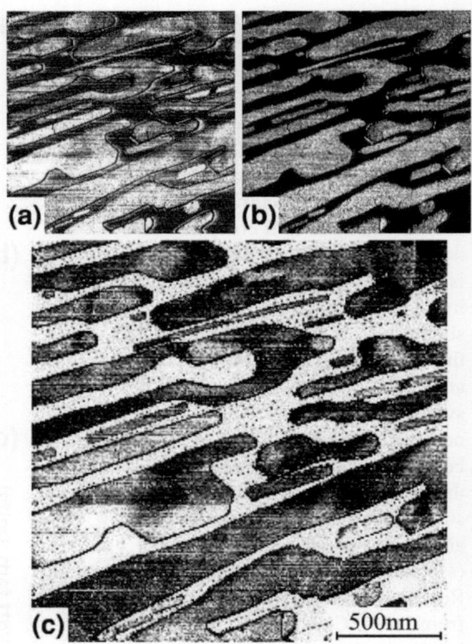

than for negative field. The opposite is true for the empty (minority) part. Free standing Gd islands on W(110) were chosen for this experiment since it is known from magneto-optical Kerr effect measurements [164] that the coercivity is only about 1.5 mT, i.e. much lower than the applied field. Therefore, one can safely conclude that the magnetization of the sample was switched by the external field while the tip magnetization remained unchanged.

The high lateral resolution down to the atomic scale is the special merit of scanning tunneling microscopy and spectroscopy. Spatially resolved measurements at $T = 70$ K with a W tip coated with approximately 10 ML Fe were performed on a sample prepared by depositing 10 ML of Gd on the W(110) substrate held at 530 K. This preparation procedure leads to partially coalesced Gd islands with a Gd wetting layer on the W(110) substrate.

The images in Fig. 5.33 were obtained from a sample in the magnetic virgin state, i.e. it was not magnetized by an external field. The scan range is 2 μm × 2 μm with 250×250 pixel^2 resolution. At every pixel a *dI/dU* spectrum was recorded. In Fig. 5.33 *dI/dU* images measured at (a) $U = -0.2$ V and (b) $U = +0.45$ V are shown, i.e. sample biases which correspond to filled and empty parts of the surface state, respectively. Since the surface state does not exist on the heavily strained first monolayer of Gd/W(110) its differential conductance is much lower than above fully relaxed Gd(0001). Consequently, the first ML of Gd/W(110) appears black. Besides obvious tip instabilities (stripes along the horizontal fast scan direction) which are most apparent at negative sample bias a strong contrast on the Gd islands in the *dI/dU* signal at $U = -0.2$ V

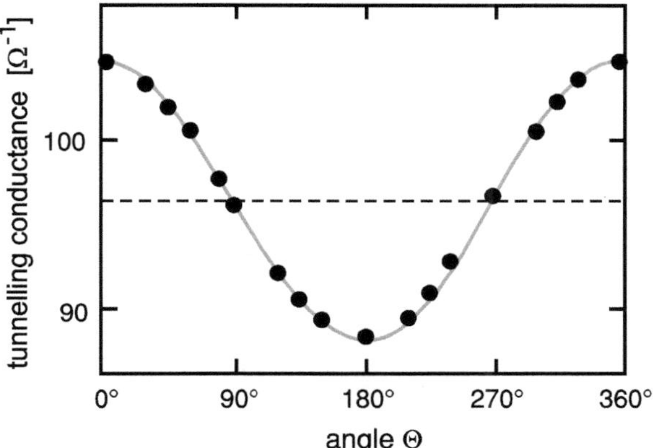

Fig. 5.34 Dependence of the tunneling conductance (inverse resistance) of a planar Fe–Al$_2$O$_3$–Fe junction on the angle Θ between the magnetization vectors of both electrodes (data taken from [165])

(see Fig. 5.33a) and a weaker contrast of opposite sign at $U = +0.45$ V (see Fig. 5.33b) is visible. In both cases the contrast originates from differences in tunneling spectra similar to those observed upon switching the sample magnetization by an external magnetic field (cf. Fig. 5.31c). The image shows not always simple light and dark contrast but intermediate. This is possible if some of the magnetic domains have a magnetization which is not collinear with the Fe spins because the differential conductance also depends on the angle Θ between both directions of magnetization. This behavior is shown in Fig. 5.34 for an Fe–Al$_2$O$_3$–Fe junction. For an arbitrary angle Θ the differential conductance can be expressed as:

$$G = G_0 \cdot (1 + P_1 P_2 \cos \Theta) \tag{5.8}$$

with G_0 being the spin averaged conductance and P_i the spin polarization of electrode i.

The influence of sporadic tip changes on the image can be reduced by introducing an asymmetry parameter A which is also used in other spin sensitive techniques in order to eliminate instrumental artifacts. Let $I_{\mathrm{maj,min}}$ be the intensity of the dI/dU signal measured at the majority (minority) peak position, then

$$A = \frac{I_{\mathrm{maj}} - I_{\mathrm{min}}}{I_{\mathrm{maj}} + I_{\mathrm{min}}}. \tag{5.9}$$

Figure 5.33c shows the asymmetry image composed by calculating A at every pixel with Figures 5.33a and 5.33b as input data. Obviously, this procedure reduces the influence of tip instabilities and enhances the domain contrast.

To evaluate the signal strength and the spatial resolution obtained so far it was zoomed into a detail, i.e. a particular Gd island already visible in the top part of Fig. 5.33. Again the dI/dU signal for both surface state spin contributions is plotted in Fig. 5.35a (filled) and 5.35(b) (empty). Both images show a domain wall crossing the island from top to bottom. A closer inspection of line sections drawn

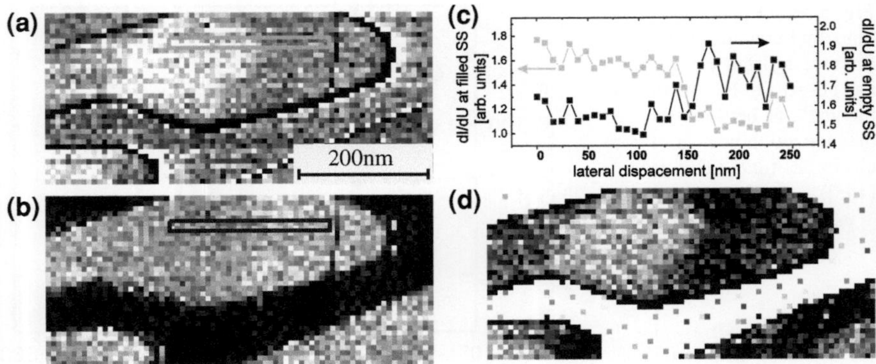

Fig. 5.35 Detail from the data set shown in Fig. 5.33. **a** In the *dI/dU* image measured at
$U = -0.2$ V the left part of the island is brighter than the right part indicating a parallel
alignment of the filled surface state spin part with the tip spin state at the Fermi level. **b** *dI/dU*
image measured at $U = +0.45$ V. **c** Plot of the *dI/dU* signal drawn along the line sections
indicated above. **d** Asymmetry image calculated from **a** and **b**. Reprinted with permission from
[130]. Copyright (1998) by the American Physical Society

from left to right along boxes across this domain wall reveals that both the filled
and the empty surface state spin part increase and decrease in intensity on a lateral
scale below 20 nm, respectively.

The correlation of magnetic properties with structure was shown for Dy thin
films on W(110), too [166]. This system was additionally investigated with
emphasis on the influence of structural defects like line defects [167] and dislocation cores [168] on the magnetic domain structure.

Figure 5.36a shows tunneling *dI/dU* spectra measured with a ferromagnetic
probe tip above adjacent domains on a Gd(0001) island. Again, the asymmetry
with respect to the differential tunneling conductance *dI/dU* at the surface state
peak positions is obvious. The contrast between these two domains was one of the
highest ever observed thus indicating that the magnetization of the tip is more or
less collinear with the magnetization of the sample. The importance of an
exchange split surface state for a reliable demonstration of spin polarized
tunneling may be illustrated by the comparison with tunneling spectra measured
above a similar sample but using a pure W tip (see inset of Fig. 5.36a). Even in the
latter case spectra could be found that exhibit significant variations in the *dI/dU*
signal at both spin parts of the surface state. However, these variations are
symmetric, i.e. both spin parts are simultaneously diminished or enhanced.

In the following the spin polarization of the Gd(0001) surface state is quantified
as observed by SPSTS and compared to data obtained by spatially averaging
surface sensitive techniques. The spin polarization P as a function of the applied
sample bias U can be estimated by

$$P_{Gd}(U) = 1/P_{Fe} \frac{dI/dU_\uparrow(U) - dI/dU_\downarrow(U)}{dI/dU_\uparrow(U) + dI/dU_\downarrow(U)}, \tag{5.10}$$

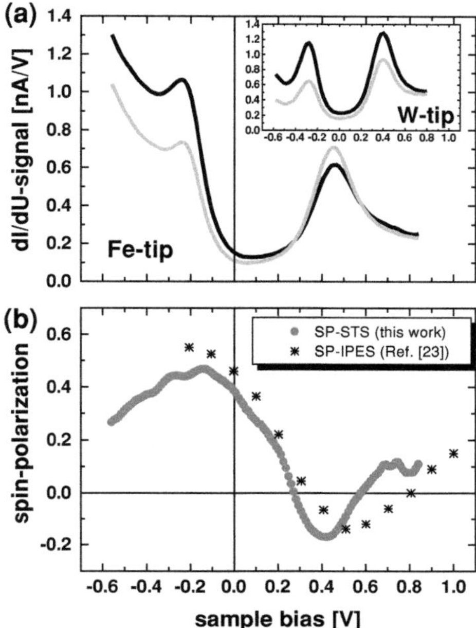

Fig. 5.36 a Tunneling spectra as measured with a Fe covered probe tip above adjacent domains. An asymmetry of the *dI/dU* signal between the empty and filled part of the surface state can clearly be recognized. In contrast, variations in the *dI/dU* signal when measured with a pure W tip are always symmetric (*inset*). **b** Spin polarization of the tunneling current between an Fe covered probe tip and the Gd(0001) surface at $T = 70\,\mathrm{K}$ (*filled circle*) compared to spin-polarized inverse photoemission data of Gd(0001) measured at $T = 130\,\mathrm{K}$ (*asterisk*) by Donath et al. [103] (reprinted with permission from [131]. Copyright 1999, American Institute of Physics)

(5.10) where $dI/dU_{\uparrow(\downarrow)}(U)$ is the intensity of the *dI/dU* signal measured above different magnetic domains and P_{Fe} is the spin polarization of the iron coated tip.

Unfortunately, there is neither a technique nor a theory available that is powerful enough to determine the spin polarization at the very end of the tip on a scale that is relevant for tunneling microscopy. In contrast, the spin polarization of thin Fe films on various substrates has consistently been determined to $P_{\mathrm{Fe}} \approx 0.4$ by different experimental methods, e.g. by means of spin polarized tunneling in planar junctions [162] or spin polarized photoemission spectroscopy [169]. For simplicity a constant spin polarization of $P_{\mathrm{Fe}} = 0.44$ [162] for the thin iron film is assumed. The dominant contribution to the bias dependence of the measured spin polarized tunnel current originates from the strong energy dependence of the spin dependent density of states of the Gd(0001) islands due to the existence of the exchange split surface state close to E_{F} whereas tunneling from or to the iron coated tip occurs via states that are essentially featureless in the relevant energy range around the Fermi level. This has been proved previously by comparing tunneling spectra from clean W(110) surfaces obtained either with clean or

Fe coated tips [170]. No additional spectroscopic feature in the energy range ± 1 eV around E_F was observed using the Fe coated tip.

It should be noted that coating of a W tip is also possible using Co [171], Gd [172], $Gd_{90}Fe_{10}$ [173], and Cr [173, 174] in order to realize SPSTM.

Application of Eq. 5.10 results in a spin polarization which behavior is shown in Fig. 5.36b. The polarization exhibits its absolutely highest values at $U = -0.13$ V and $U = +0.42$ V, i.e. at bias voltages just below the peak positions of the surface state. For comparison values of the spin polarization of homogeneous, approximately 30 ML thick Gd(0001) films grown on W(110), as determined by means of spin resolved inverse photoemission (SPIPE) [103], have been enclosed. An excellent overall agreement can be recognized. Both, SPSTS as well as SPIPE data exhibit a positive spin polarization P_{Gd} on both sides of the Fermi level. In more detail, P_{Gd} amounts to about 0.5 close to the position of the majority surface state, decreases to $P_{Gd} \approx 0.4$ at the Fermi level, and vanishes about 300 meV above the the Fermi level. With both experimental techniques a negative spin polarization can only be recognized near the peak position of the minority surface state, while the spin polarization turns to positive values again for unoccupied electronic states being energetically further apart from the Fermi level.

Obviously, a magnetically dependent signal measured with a highly surface sensitive technique like STM and STS must decay with time elapsed from the moment of surface preparation. In order to estimate the time dependence of the spin polarization of the tunneling current the same sample was measured in three subsequent scans. A similar procedure was used to proof the surface sensitivity of optical second harmonic generation [175]. Since the measurement was performed by taking a dI/dU spectrum at every pixel a single scan requires approximately 24 h. Three locations of the sample showing strong dark/bright contrasts separated by domain walls were analyzed in detail. The spin polarization at bias voltages that correspond to the peak positions of the Gd(0001) surface state was calculated by using Eq. 5.10. It is immediately apparent from the semi-logarithmic plot shown in Fig. 5.37 that the spin polarization decreases from scan to scan. Assuming an exponential decay one finds by regression analysis a decay constant $\tau^{maj} = 81$ h and $\tau^{min} = 32$ h and an initial spin polarization $P_0^{maj} = 0.48$ and $P_0^{min} = 0.20$ for the majority and minority spin part of the surface state, respectively. Although the error bars are rather large especially in the case of the minority spin part (approximately $\pm 15\%$) the initial spin polarization values are in good agreement with temperature dependent spin resolved (inverse) PES data [103, 121]. Surprisingly, a difference in the decay constant as obtained for the majority and minority part of the surface state is observed.

The influence of the stray field of the probe tip on the domain structure of the sample is a critical parameter which has to be considered in order to determine the significance of the experimental images. No indication of tip induced changes of the sample domain structure for Fe coatings with $\Theta_{tip} \leq 10$ ML was ever found; if, however, $\Theta_{tip} \geq 100$ ML tip induced magnetic modifications were observed. Figure 5.38 shows the asymmetry image of Gd(0001) islands grown on a W(110) substrate. The sample was scanned from bottom to top with a tip coated by an

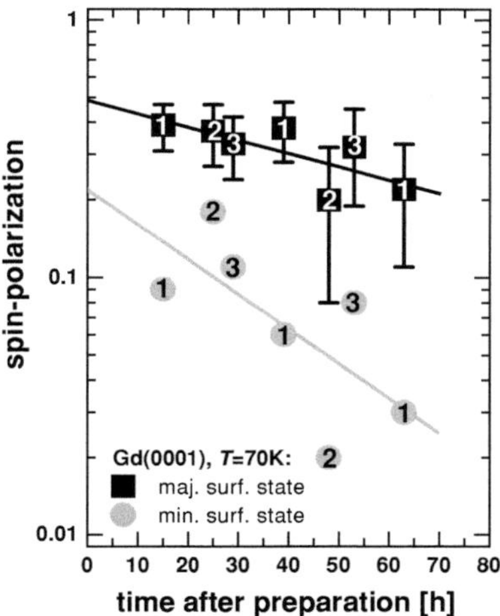

Fig. 5.37 Plot of the spin polarization of the tunneling current at bias voltages that correspond to the binding energies of the majority and minority part of the Gd(0001) surface state (the ordinate is on a logarithmic scale) as obtained by measuring the same sample in three subsequent scans during a time period of approximately 24 h. The data points marked *1*, *2*, and *3* denote three different locations of the sample. Apparently, the spin polarization decreases with time elapsed from surface preparation (reprinted with permission from [131]. Copyright 1999, American Institute of Physics)

Fig. 5.38 Asymmetry image of a Gd sample measured with an approximately 150 ML thick Fe coating on the tip. The *circles* indicate two Gd islands whose magnetization were accidentally switched probably due to interaction between tip and sample mediated via the stray field. Reprinted with permission from [130]. Copyright (1998) by the American Physical Society

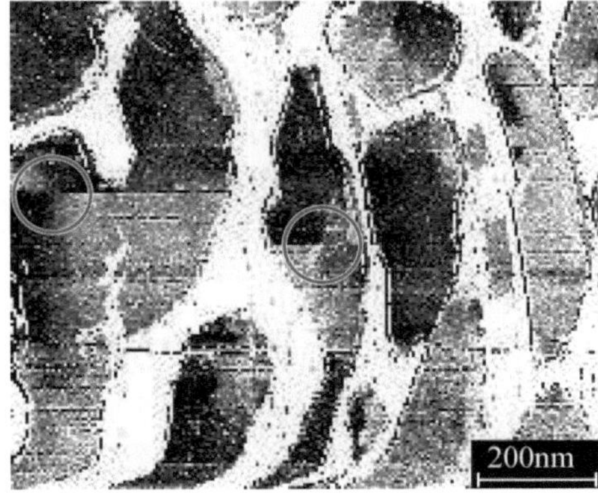

Fe film of approximately 150 ML thickness. Two circles indicate locations where a checkerboard pattern was observed which is characteristic for a magnetization reversal of single Gd islands. Magnetic switching of the tip can be ruled out since other Gd islands are imaged unchanged in subsequent scan lines. In order to overcome this influence of the tip it is possible to use a tip consisting of antiferromagnetic material like MnNi [176] or to coat the non-magnetic probe tips with antiferromagnetic films like Cr [173, 174].

A comprehensive overview on SPSTS concerning the magnetic spin structure with atomic resolution, of nanoscaled wires, nanoscaled elements with magnetic vortex structures, and chiral magnetic order is given in [143].

References

1. G. Binnig, H. Rohrer, Rev. Mod. Phys. **59**, 615 (1987)
2. M. Getzlaff, *Fundamentals of Magnetism* (Springer, Berlin, 2008)
3. C. Powell, J. Electr. Spectr. Relat. Phen. **47**, 197 (1988)
4. M. Seah, W. Dench, Surf. Interface Anal. **1**, 2 (1979)
5. J. Bansmann, M. Getzlaff, C. Westphal, G. Schönhense, J. Magn. Magn. Mater. **117**, 38 (1992)
6. M. Landolt, Appl. Phys. A **41**, 83 (1986)
7. D. Venus, J. Kirschner, Phys. Rev. B **37**, 2199 (1988)
8. D. Pappas, K.P. Kämper, B. Miller, H. Hopster, D. Fowler, C. Brundle, A. Luntz, Z.X. Shen, Phys. Rev. Lett. **66**, 504 (1991)
9. M. Getzlaff, J. Bansmann, G. Schönhense, Sol. State Comm. **87**, 467 (1993)
10. M. Getzlaff, J. Bansmann, C. Westphal, G. Schönhense, J. Magn. Magn. Mater. **104–107**, 1781 (1992)
11. V. Karpan, G. Giovannetti, P. Khomyakov, M. Talanana, A. Starikov, M. Zwierzycki, J. van den Brink, G. Brocks, P. Kelly, Phys. Rev. Lett. **99**, 176602 (2007)
12. Y. Dedkov, M. Fonin, C. Laubschat, Appl. Phys. Lett. **92**, 052506 (2008)
13. M. Getzlaff, J. Bansmann, J. Braun, G. Schönhense, Z. Phys. B **104**, 11 (1997)
14. M. Getzlaff, J. Bansmann, J. Braun, G. Schönhense, J. Magn. Magn. Mater. **161**, 70 (1996)
15. K. Schröder, G. Prinz, K.H. Walker, E. Kisker, J. Appl. Phys. **57**, 3669 (1985)
16. A. Turner, J. Erskine, Phys. Rev. B **25**, 1983 (1982)
17. J. Callaway, C. Wang, Phys. Rev. B **16**, 2095 (1977)
18. J. Redinger, C. Fu, A. Freeman, U. König, P. Weinberger, Phys. Rev. B **38**, 5203 (1988)
19. E. Kisker, K. Schröder, M. Campagna, W. Gudat, Phys. Rev. Lett. **52**, 2285 (1984)
20. E. Kisker, K. Schröder, W. Gudat, M. Campagna, Phys. Rev. B **31**, 329 (1985)
21. H. Danan, A. Herr, A. Meyer, J. Appl. Phys. **39**, 669 (1968)
22. F. Himpsel, J. Magn. Magn. Mater. **102**, 261 (1991)
23. Y. Sakisaka, T. Maruyama, H. Kato, Y. Aiura, H. Yanashima, Phys. Rev. B **41**, 11865 (1990)
24. R. Kurzawa, Ph.D. thesis, Universität Köln (1986)
25. M. Getzlaff, B. Heidemann, J. Bansmann, C. Westphal, G. Schönhense, Rev. Sci. Instr. **69**, 3913 (1998)
26. J. Bansmann, M. Getzlaff, G. Schönhense, M. Fluchtmann, J. Braun, Surf. Sci. **402–404**, 365 (1998)
27. J. Bansmann, M. Getzlaff, C. Ostertag, G. Schönhense, Surf. Sci. **352–354**, 898 (1996)
28. J. Bansmann, L. Lu, M. Getzlaff, M. Fluchtmann, J. Braun, J. Magn. Magn. Mater. **185**, 94 (1998)
29. K. Hathaway, H. Jansen, A. Freeman, Phys. Rev. B **31**, 7603 (1985)
30. L. Baumgarten, C. Schneider, H. Petersen, F. Schäfers, J. Kirschner, Phys. Rev. Lett. **65**, 492 (1990)
31. H. Rose, C. Roth, F. Hillebrecht, E. Kisker, Sol. State Comm. **91**, 129 (1994)
32. N. Cherepkov, Phys. Rev. B **50**, 13813 (1994)
33. D. Venus, Phys. Rev. B **49**, 8821 (1994)
34. B. Thole, G. van der Laan, Phys. Rev. B **44**, 12424 (1991)
35. H. Ebert, L. Baumgarten, C. Schneider, J. Kirschner, Phys. Rev. B **44**, 4406 (1991)

36. U. Fano, Phys. Rev. **178**, 131 (1969)
37. J. Kessler, *Polarized Electrons*, 2nd edn. (Springer, Berlin, 1985)
38. N. Cherepkov, Sov. Phys. – JETP **38**, 463 (1974)
39. U. Heinzmann, G. Schönhense, J. Kessler, Phys. Rev. Lett. **42**, 1603 (1979)
40. G. Schönhense, Phys. Rev. Lett. **44**, 640 (1980)
41. N. Cherepkov, V. Kuznetsov, J. Phys. B **22**, L405 (1989)
42. M. Getzlaff, C. Ostertag, G. Fecher, N. Cherepkov, G. Schönhense, Phys. Rev. Lett. **73**, 3030 (1994)
43. K. Blum, *Density Matrix Theory and Applications*. (Plenum Press, New York, 1981)
44. J. Gadzuk, Phys. Rev. B **10**, 5030 (1974)
45. G. Baum, D. Egert, M. Getzlaff, W. Raith, H. Steidl, J. Magn. Magn. Mater. **148**, 179 (1995)
46. K. Yu, W. Spicer, I. Lindau, P. Pianetta, S. Lin, Surf. Sci. **57**, 157 (1976)
47. W. Sesselmann, B. Woratschek, J. Küppers, G. Ertl, H. Haberland, Phys. Rev. B **35**, 1547 (1987)
48. M. Hammond, F. Dunning, G. Walters, G. Prinz, Phys. Rev. B **45**, 3674 (1992)
49. M. Salvietti, P. Ferro, R. Moroni, M. Canepa, L. Mattera, Surf. Sci. **377–379**, 481 (1997)
50. R. Wu, A. Freeman, Phys. Rev. Lett. **69**, 2867 (1992)
51. A. Freeman, C. Fu, J. Appl. Phys. **61**, 3356 (1987)
52. J. Bansmann, L. Lu, M. Getzlaff, K. Meiwes-Broer, Z. Phys. D **40**, 570 (1997)
53. J. Bansmann, L. Lu, M. Getzlaff, K. Meiwes-Broer, Surf. Sci. **377–379**, 476 (1997)
54. J. Bandmann, L. Lu, M. Getzlaff, K. Meiwes-Broer, in *Small Particles and Clusters*, ed. by H. Andersen (Springer, Berlin, 1997)
55. J. Bansmann, L. Lu, V. Senz, A. Bettac, M. Getzlaff, K. Meiwes-Broer, Eur. Phys. J. D 9:461
56. M. Bode, R. Pascal, R. Wiesendanger, J. Vac. Sci. Technol. A **15**, 1285 (1997)
57. G. van der Laan, Phys. Rev. B **51**, 240 (1995)
58. J. Ociepa, P. Schultz, K. Griffiths, P. Norton, Surf. Sci. **225**, 281 (1990)
59. M. Bode, R. Pascal, R. Wiesendanger, Phys. Bl. **52**, 551 (1996)
60. W. Kuch, M.T. Lin, K. Meinel, C. Schneider, J. Noffke, J. Kirschner, Phys. Rev. B **51**, 12627 (1995)
61. J. Bansmann, L. Lu, M. Getzlaff, M. Fluchtmann, J. Braun, Surf. Sci. **454**, 686 (2000)
62. H. Knoppe, E. Bauer, Phys. Rev. B **48**, 1794 (1993)
63. G. Garreau, M. Farle, E. Beaurepaire, K. Baberschke, Phys. Rev. B **55**, 330 (1997)
64. Y. Sakisaka, T. Komeda, T. Miyano, M. Onchi, S. Masuda, Y. Harada, K. Yagi, H. Kato, Surf. Sci. **164**, 220 (1985)
65. M. Getzlaff, J. Bansmann, G. Schönhense, J. Magn. Magn. Mater. **192**, 458 (1999)
66. H. Huang, J. Hermanson, Phys. Rev. B **32**, 6312 (1985)
67. S. Masuda, Y. Harada, H. Kato, K. Yagi, T. Komeda, T. Miyano, M. Onchi, Y. Sakisaka, Phys. Rev. B **37**, 8088 (1988)
68. T. Miyano, Y. Sakisaka, T. Komeda, M. Onchi, Surf. Sci. **169**, 197 (1986)
69. M. Getzlaff, J. Bansmann, G. Schönhense, Fresenius J. Anal. Chem. **353**, 748 (1995)
70. M. Getzlaff, G. Schönhense, Surf. Sci. **331–333**, 213 (1995)
71. M. Getzlaff, J. Bansmann, G. Schönhense, Fresenius J. Anal. Chem. **353**, 743 (1995)
72. S. Chubb, W. Pickett, Phys. Rev. Lett. **58**, 1248 (1987)
73. S. Chubb, W. Pickett, Sol. State Comm. **62**, 19 (1987)
74. R. Wu, A. Freeman, J. Appl. Phys. **73**, 6739 (1993)
75. T. Walker, H. Hopster, Phys. Rev. B **48**, 3563 (1993)
76. W. Weber, D. Wesner, D. Hartmann, G. Güntherodt, Phys. Rev. B **46**, 6199 (1992)
77. F. Himpsel, Phys. Rev. Lett. **67**, 2363 (1991)
78. M. Getzlaff, J. Bansmann, G. Schönhense, J. Magn. Magn. Mater. **131**, 304 (1994)
79. M. Bridge, R. Lambert, Surf. Sci. **82**, 413 (1979)
80. G. Castro, J. Küppers, Surf. Sci. **123**, 456 (1982)
81. F. Grellner, B. Klingenberg, D. Borgmann, G. Wedler, J. Electr. Spectr. Relat. Phen. **71**, 107 (1995)

82. Y. Jugnet, T. Duc, J. Phys. Chem. Solids **40**, 29 (1979)
83. H.J. Freund, G. Hohlneicher, Ber. Bunsenges. Phys. Chem. **83**, 100 (1979)
84. S. Hüfner, in *Photoemission in Solids II*, chap. 3, ed. by L. Ley, M. Cardona (Springer, Berlin, 1979)
85. M. Getzlaff, J. Bansmann, G. Schönhense, J. Magn. Magn. Mater. **140–144**, 729 (1995)
86. M. Getzlaff, J. Bansmann, G. Schönhense, J. Electr. Spectr. Relat. Phen. **77**, 197 (1996)
87. W. Clemens, E. Vescovo, T. Kachel, C. Carbone, W. Eberhardt, Phys. Rev. B **46**, 4198 (1992)
88. M. Getzlaff, J. Bansmann, G. Schönhense, Phys. Lett. A **182**, 153 (1993)
89. M. Getzlaff, J. Bansmann, G. Schönhense, J. Chem. Phys. **103**, 6691 (1995)
90. M. Getzlaff, J. Bansmann, G. Schönhense, Surf. Sci. **323**, 118 (1995)
91. M. Getzlaff, J. Bansmann, G. Schönhense, Phys. Rev. Lett. **71**, 793 (1993)
92. M. Getzlaff, N. Cherepkov, G. Schönhense, Phys. Rev. B **52**, 3421 (1995)
93. J. Bansmann, L. Lu, M. Getzlaff, Surf. Sci. **402–404**, 371 (1998)
94. G. Panaccione, F. Sirotti, G. Rossi, J. Electr. Spectr. Relat. Phen. **76**, 189 (1995)
95. M. Getzlaff, D. Egert, P. Rappolt, M. Wilhelm, H. Steidl, G. Baum, W. Raith, J. Magn. Magn. Mater. **140–144**, 727 (1995)
96. M. Getzlaff, D. Egert, P. Rappolt, M. Wilhelm, H. Steidl, G. Baum, W. Raith, Surf. Sci. **331–333**, 1404 (1995)
97. M. Onellion, M. Hart, F. Dunning, G. Walters, Phys. Rev. Lett. **52**, 380 (1984)
98. D. Weller, S. Alvarado, W. Gudat, K. Schröder, M. Campagna, Phys. Rev. Lett. **54**, 1555 (1985)
99. D. Weller, S. Alvarado, J. Appl. Phys. **59**, 2908 (1986)
100. R. Wu, C. Li, A. Freeman, C. Fu, Phys. Rev. B **44**, 9400 (1991)
101. D. Li, C. Hutchings, P. Dowben, C. Hwang, R.T. Wu, M. Onellion, A. Andrews, J. Erskine, J. Magn. Magn. Mater. **99**, 85 (1991)
102. D. Li, P. Dowben, J. Ortega, F. Himpsel, Phys. Rev. B **49**, 7734 (1994)
103. J. Quinn, Y. Li, F. Jona, D. Fort, Phys. Rev. B **46**, 9694 (1992)
104. M. Donath, B. Gubanka, F. Passek, Phys. Rev. Lett. **77**, 5138 (1996)
105. G. Mulhollan, K. Garrison, J. Erskine, Phys. Rev. Lett. **69**, 3240 (1992)
106. K. Starke, E. Navas, L. Baumgarten, G. Kaindl, Phys. Rev. B **48**, 1329 (1993)
107. D. Bylander, L. Kleinman, Phys. Rev. B **49**, 1608 (1994)
108. E. Stoner, Proc. R. Soc. A **154**, 656 (1936)
109. E. Wohlfarth, Rev. Mod. Phys. **25**, 211 (1953)
110. V. Korenman, J. Murray, R. Prange, Phys. Rev. B **16**, 4032 (1977)
111. V. Korenman, J. Murray, R. Prange, Phys. Rev. B **16**, 4048 (1977)
112. V. Korenman, J. Murray, R. Prange, Phys. Rev. B **16**, 4058 (1977)
113. A. Kakizaki, J. Fujii, K. Shimada, A. Kamata, K. Ono, K.H. Park, T. Kinoshita, T. Ishii, H. Fukutani, Phys. Rev. Lett. **72**, 2781 (1994)
114. J. Kirschner, M. Glöbl, V. Dose, H. Scheidt, Phys. Rev. Lett. **53**, 612 (1984)
115. P. Aebi, T. Kreutz, J. Osterwalder, R. Fasel, P. Schwaller, L. Schlapbach, Phys. Rev. Lett. **76**, 1150 (1996)
116. E. Weschke, C. Schüssler-Langeheine, R. Meier, A. Fedorov, K. Starke, F. Hübinger, G. Kaindl, Phys. Rev. Lett. **77**, 3415 (1996)
117. A. Fedorov, K. Starke, G. Kaindl, Phys. Rev. B **50**, 2739 (1994)
118. B. Kim, A. Andrews, J. Erskine, K. Kim, B. Harmon, Phys. Rev. Lett. **68**, 1931 (1992)
119. M. Getzlaff, M. Bode, S. Heinze, R. Pascal, R. Wiesendanger, J. Magn. Magn. Mater. **184**, 155 (1998)
120. M. Bode, M. Getzlaff, S. Heinze, R. Pascal, R. Wiesendanger, Appl. Phys. A **66**:S121 (1998)
121. D. Li, J. Pearson, S. Bader, D. McIlroy, C. Waldfried, P. Dowben, Phys. Rev. B **51**, 13895 (1995)
122. F. Hübinger, C. Schüssler-Langeheine, A. Fedorov, K. Starke, E. Weschke, A. Höhr, S. Vandrè, G. Kaindl, J. Electr. Spectr. Relat. Phen. **76**, 535 (1995)

123. H. Tang, D. Weller, T. Walker, J. Scott, C. Chappert, H. Hopster, A. Pang, D. Dessau, D. Pappas, Phys. Rev. Lett. **71**, 444 (1993)
124. C. Rau, C. Jin, M. Robert, Phys. Lett. A **138**, 334 (1989)
125. S. Wu, H. Li, Y. Li, D. Tian, J. Quinn, F. Jona, D. Fort, Phys. Rev. B **44**, 13720 (1991)
126. R. Ahuja, S. Auluck, B. Johansson, M. Brooks, Phys. Rev. B **50**, 5147 (1994)
127. L. Sandratskii, J. Kübler, Europhys. Lett. **23**, 661 (1993)
128. M. Bode, M. Getzlaff, R. Pascal, S. Heinze, R. Wiesendanger, in *Magnetism and Electronic Correlations in Local-moment Systems: Rare Earth Elements and Compounds*, ed. by M. Donath, P. Dowben, W. Nolting (World Scientific, Singapore, 1998)
129. M. Getzlaff, M. Bode, S. Heinze, R. Wiesendanger, Appl. Surf. Sci. **142**, 558 (1999)
130. M. Bode, R. Pascal, M. Getzlaff, R. Wiesendanger, Acta Phys. Pol. **93**, 273 (1999)
131. M. Bode, M. Getzlaff, R. Wiesendanger, Phys. Rev. Lett. **81**, 4256 (1998)
132. M. Bode, M. Getzlaff, R. Wiesendanger, J. Vac. Sci. Technol. A **17**, 2228 (1999)
133. R. Wiesendanger, M. Bode, M. Getzlaff, Appl. Phys. Lett. **75**, 124 (1999)
134. M. Bode, M. Getzlaff, A. Kubetzka, R. Pascal, O. Pietzsch, R. Wiesendanger, Phys. Rev. Lett. **83**, 3017 (1999)
135. M. Farle, K. Baberschke, U. Stetter, A. Aspelmeier, F. Gerhardter, Phys. Rev. B **47**, 11571 (1993)
136. R. Greenough, N. Hettiarachchi, D. Jones, J. Hulbert, J. Magn. Magn. Mater. **29**, 67 (1982)
137. H. Nigh, S. Legvold, F. Spedding, Phys. Rev. **132**, 1092 (1963)
138. E. Tober, F. Palomares, R. Ynzunza, R. Denecke, J. Morais, Z. Wang, G. Bino, J. Liesegan, Z. Hussain, C. Fadley, Phys. Rev. Lett. **81**, 2360 (1998)
139. H. Schröder, Nucl. Instrum. Meth. **194**, 381 (1982)
140. M. Bode, M. Dreyer, M. Getzlaff, M. Kleiber, A. Wadas, R. Wiesendanger, J. Phys. Condens. Matter **11**, 9387 (1999)
141. M. Getzlaff, M. Bode, R. Wiesendanger, Surf. Rev. Lett. **6**, 591 (1999)
142. R. Wiesendanger, M. Bode, M. Getzlaff, J. Magn. Soc. Jpn. **23**, 195 (2000)
143. M. Getzlaff, in *Nanotechnology*, ed. by H. Fuchs, Nanoprobes, Chap. 1, vol. 6. (Wiley-VCH, Weinheim, 2009)
144. J. Slonczewski, Phys. Rev. B **39**, 6995 (1989)
145. M. Julliere, Phys. Lett. A **54**, 225 (1975)
146. S. Maekawa, U. Gäfvert, IEEE Trans. Mag. **18**, 707 (1982)
147. J. Moodera, L. Kinder, T. Wong, R. Meservey, Phys. Rev. Lett. **74**, 3273 (1995)
148. J. Moodera, L. Kinder, J. Appl. Phys. **79**, 4724 (1996)
149. R. Wiesendanger, H.J. Güntherodt, G. Güntherodt, R. Gambino, R. Ruf, Phys. Rev. Lett. **65**, 247 (1990)
150. W. Wulfhekel, J. Kirschner, Appl. Phys. Lett. **75**, 1944 (1999)
151. W. Wulfhekel, H. Ding, J. Kirschner, J. Appl. Phys. **87**, 6475 (2000)
152. W. Wulfhekel, H. Ding, W. Lutzke, G. Steierl, M. Vázquez, P. Marin, A. Hernando, J. Kirschner, Appl. Phys. A **72**, 463 (2001)
153. H. Ding, W. Wulfhekel, C. Chen, J. Barthel, J. Kirschner, Mater. Sci. Eng. B **84**, 96 (2001)
154. H. Ding, W. Wulfhekel, J. Kirschner, Europhys. Lett. **57**, 100 (2002)
155. W. Wulfhekel, R. Hertel, H. Ding, G. Steierl, J. Kirschner, J. Magn. Magn. Mater. **249**, 368 (2002)
156. H. Ding, W. Wulfhekel, U. Schlickum, J. Kirschner, Europhys. Lett. **63**, 419 (2003)
157. U. Schlickum, W. Wulfhekel, J. Kirschner, Appl. Phys. Lett. **83**, 2016 (2003)
158. U. Schlickum, N. Janke-Gilman, W. Wulfhekel, J. Kirschner, Phys. Rev. Lett. **92**, 107203 (2004)
159. S. Alvarado, P. Renaud, Phys. Rev. Lett. **68**, 1387 (1992)
160. S. Alvarado, Phys. Rev. Lett. **75**, 513 (1995)
161. P. Tedrow, R. Meservey, P. Fulde, Phys. Rev. Lett. **25**, 1270 (1970)
162. P. Tedrow, R. Meservey, Phys. Rev. B **7**, 318 (1973)
163. J. Stroscio, D. Pierce, A. Davies, R. Celotta, M. Weinert, Phys. Rev. Lett. **75**, 2960 (1995)
164. M. Farle, W. Lewis, J. Appl. Phys. **75**, 5604 (1994)

165. T. Miyazaki, N. Tezuka, J. Magn. Magn. Mater. **139**, L231 (1995)
166. L. Berbil-Bautista, S. Krause, M. Bode, R. Wiesendanger, Phys. Rev. B 76:064411 (2007)
167. S. Krause, L. Berbil-Bautista, T. Hänke, F. Vonau, M. Bode, R. Wiesendanger, Europhys. Lett. **76**, 637 (2006)
168. L. Berbil-Bautista, S. Krause, M. Bode, A. Badia-Majós, C. de la Fuente, R. Wiesendanger, J. Arnaudas, Phys. Rev. B **80**, 241408 (2009)
169. B. Sinković, E. Shekel, S. Hulbert, Phys. Rev. B **52**, R8696 (1995)
170. R. Wiesendanger, MRS Bull. **22**, 31 (1997)
171. J. Prokop, A. Kukunin, H. Elmers, Phys. Rev. B **73**, 014428 (2006)
172. O. Pietzsch, A. Kubetzka, M. Bode, R. Wiesendanger, Phys. Rev. Lett. **84**, 5212 (2000)
173. A. Kubetzka, M. Bode, O. Pietzsch, R. Wiesendanger, Phys. Rev. Lett. **88**, 057201 (2002)
174. A. Wachowiak, J. Wiebe, M. Bode, O. Pietzsch, M. Morgenstern, R. Wiesendanger, Science **298**, 577 (2002)
175. J. Reif, J. Zink, C. Schneider, J. Kirschner, Phys. Rev. Lett. **67**, 2878 (1991)
176. S. Murphy, J. Osing, I. Shvets, Appl. Surf. Sci. **144–145**, 497 (1999)

Chapter 6
Summary

At the beginning the important aspects concerning the experimental techniques which were used to be sensitive to surface magnetic properties were discussed. It was explained why in photoemission techniques information can be obtained by determination of the electron spin polarization or by using polarized light for the excitation process. An enhancement of the surface sensitivity is obtained by investigation of tunneling electrons. The related techniques were described, scanning tunneling microscopy with its high lateral resolution and in more detail the de-excitation mechanisms of excited spin polarized noble gas atoms. Both tools were discussed in terms of being sensitive to magnetic properties.

The relationship between structural and electronic, thus also magnetic, behavior of thin film rare earth metal systems was investigated using spatial averaging techniques as well as such with a high lateral resolution. It turned out that for gadolinium and terbium a thickness of four layers is sufficient to observe the fully developed exchange split surface state. Using the spectroscopic mode of the scanning tunneling microscope (STM) the detection of stacking faults was realized. For the binary alloy $GdFe_2$ it was demonstrated that it is possible to obtain an epitaxial growth resulting in the thin film system to be in a single crystalline state. Ternary intermetallics with rare earth metals as one constituent exhibiting extraordinary properties, so-called Heavy Fermion Systems, were prepared to exhibit atomically clean and well-ordered surfaces thus being accessible to electron spectroscopy. Using resonant photoemission the rare earth metal related features were shown to be separated from those of the other constituents.

The "chemistry" of a surface determines the surface electronic properties. The following chapter was therefore focused on the influence of adsorbates. It was found that hydrogen exhibit unusual adsorption characteristics demonstrating a lagoon-like appearance. The incorporation of hydrogen in Gd thin films leads to a plastic deformation resulting in surface modifications which were identified by STM. The combination of ultraviolet photoelectron spectroscopy (UPS) and STM allowed to determine the reaction scheme of coadsorption processes, exemplarily presented for hydrogen covered Gd surfaces exposed to CO.

M. Getzlaff, *Surface Magnetism*, Springer Tracts in Modern Physics, 240,
DOI: 10.1007/978-3-642-14189-8_6, © Springer-Verlag Berlin Heidelberg 2010

This knowledge of structural and electronic properties is the indispensable basis for the final and main part concentrated on surface magnetic properties. The determination of electronic band structure effects like an exchange splitting was exemplarily demonstrated for Fe(110) thin films. Different photoemission techniques were shown to give complementary information concerning the band structure. Magnetic dichroism in photoemission allowed to compare the magnetic behavior like spin reorientation phenomena of epitaxially grown flat films and nanostructured systems. Non-magnetic adsorbate atoms on ferromagnetic surfaces were shown to influence the magnetic properties of the substrate. Additionally, it could be demonstrated that such an atom becomes magnetic itself. Using ferromagnetic probe tips in STM experiments allows to realize spin polarized vacuum tunneling. This new technique was applied to image magnetic domains on the nanometer scale. Additionally, it could be demonstrated that the stray field of the tip can be used to switch the magnetization of nanoscale islands.

Index